MAKING UP
NUMBERS

Making up Numbers

A History of Invention in Mathematics

Ekkehard Kopp

ISBN Paperback: 978-1-80064-095-5

ISBN Hardback: 978-1-80064-096-2

ISBN Digital (PDF): 978-1-80064-097-9

DOI: 10.11647/OBP.0236

Cover images from Wikimedia Commons. For image details see captions in the book.

Cover design by Anna Gatti.

Contents

For Marianne

Preface

Human beings have an innate need to make things up. People make up stories, nations make up histories, scientists make up theories to explain how the world works and philosophers ponder how we know things and how we should live and behave. These made-up tales often conflict with each other, but perhaps there is one thing on which we can all agree: that it is necessary to *make up numbers* to help us cope with life and with each other, from times when 'one, two, many' seemed to be enough, right down to the modern concepts of number used by scientists and mathematicians today. We might not always agree, nor even think about, what numbers *are*, but no-one is likely to deny that we need them.

Numbers crop up everywhere in modern life: on clocks, calendars, coins and in cash dispensers, for example. At primary school we all spent much time learning to manipulate numbers: we added and subtracted, learnt multiplication tables by rote, practised long division—some of us even learnt how to compute square roots. Much of this is now done routinely with calculators and computers and we forget the effort spent in acquiring the basics when we were young—perhaps we even forget how to use them.

If you have ever wondered how all this came about, how our concept of numbers has developed over the centuries, and how various puzzles and conceptual problems encountered along the way were resolved, then this book should be of interest to you. You might be a current or intending mathematics undergraduate, or a keen student of A-level mathematics, or indeed be teaching the subject at secondary school. Or you might simply be interested in mathematics and seek to learn more about its development.

The traditional mathematics syllabus, at school, college or university, at best makes passing reference to the fascinating history of our subject. Students seeking to trace the development of mathematical ideas often find that there are relatively few detailed but accessible sources to guide them; and while texts presenting 'popular mathematics' can provide much fun with examples and interesting anecdotes, the thread of conceptual development sometimes suffers in the process.

This book makes no pretence to be an academic treatise in the history of mathematics, nor is it a mathematics textbook. It seeks to tell a story, one that I hope may inform readers whose prior experience of abstract mathematical arguments is not extensive.

To understand what mathematics does and how it has developed, it is essential to *do* some mathematics. In presenting problems whose solutions led to ever wider classes of number, as well as discussing conceptual obstacles that were overcome, I make use of mathematical notation, basic manipulation of equations and step-by-step mathematical reasoning. Some of this has been placed in shaded sections that readers in a hurry may decide to skip, hopefully without loss of continuity. To assist readers seeking more detail on particular points, an online resource—available at https://www.openbookpublishers.com/product/1279#resources—entitled *Mathematical Miscellany* (abbreviated to *MM* in the text) accompanies this book. Its purpose is to remind the reader of basic mathematical concepts, provide simple technical details, as well as some longer proofs, that are omitted in the text, and provide more background, mathematical and historical, on topics addressed in the book.

It may seem that nothing more needs to be said about numbers. So it may surprise some readers of the final chapters that mathematicians today are not immune to doubts about the foundations of their subject. After all, the rigour of mathematical proof and the timelessness of mathematical truths have been hallmarks of the discipline ever since Ancient Greece, more than 2000 years ago. Until quite recently, countless generations of school pupils spent years wrestling with the inexorable logic of the geometric constructions and theorems in Euclid's *Elements*. Today they also encounter the abstraction of algebraic symbols in solving equations and (somewhat later) marvel at the apparently miraculous success of the Calculus in the quantitative analysis of motion and forces in our physical universe, which led, in turn, to technological revolutions that now govern our everyday lives. Why, indeed, should any of this be subject to doubt?

Naturally, I am not claiming that I am beset with doubt. Rather, I regard mathematics as a human activity, whose historical development reflects the continuing refinement and abstraction of its concepts—including the concept of number, and even that of proof—as a process of evolution. This process is conducted collectively and is stimulated by careful observation of our environment, creative use of the imagination, and intellectual rigour. From that perspective it does not seem so different from other human endeavours. It is not infallible, nor are its precepts beyond question, however well-hidden or abstruse they may be. In the final chapter of this book this is illustrated, in a graphic account of disputes over the foundations of the subject, by the eminent mathematician John von Neumann, who, over seventy years ago, explained the conundrum posed there more vividly than I can today.

Viewed in this light, the lives, work, achievements and strivings of mathematicians, ancient and modern, might perhaps be seen in more human terms. Those who teach the subject, at any level, might find such historical perspectives helpful when seeking to overcome the all too prevalent perception of the subject as 'too difficult', or even as 'dry' and devoid of human drama, humour or fallibility.

Acknowledgements:

I have endeavoured, in the Bibliography and in various footnotes, to identify the sources (which, in the main, are admittedly secondary) from which the material developed in this book was taken. I apologise for any omissions in attribution and make no claims of originality for any mathematical ideas presented here.

Friends and colleagues have kindly given their time to help make the content more accurate and comprehensible: Howell Lloyd encouraged me to address the subject matter from a historical perspective; Dona Strauss read an earlier draft, making many invaluable suggestions; the current text benefits from helpful advice by Tomasz Zastawniak, while Nigel Cutland was generous in his careful reading of the text and sharing his far greater expertise in logic; Tony Gardiner provided numerous thoughtful comments and corrections as well as suggesting OBP; Maciek Capinski was a constant and invaluable source of expertise and support in overcoming my struggles with LaTeX and greatly improved the graphics. My sincere thanks go to them all.

I am grateful to an anonymous referee for insightful comments, and to Alessandra Tosi, Melissa Purkiss, Anna Gatti and Luca Baffa at Open Book Publishers for their unfailingly helpful cooperation and their expertise in bringing this project to a successful conclusion.

All remaining errors and misconceptions are entirely my own.

My dear wife, Margaret, has patiently endured my pre-occupations and frustrations for much longer than was originally envisaged, and has provided comfort and constant support. This project is older than my young granddaughter Marianne—the book is dedicated to her in the hope that she may enjoy it one day.

.

Prologue: Naming Numbers

In the mathematics I can report no deficience, except it be that men do not sufficiently understand this excellent use of the pure mathematics, in that they do remedy and cure many defects in the wit and faculties intellectual. For if the wit be too dull, they sharpen it; if too wandering, they fix it; if too inherent in the sense, they abstract it.

Sir Francis Bacon, *The Advancement of Learning*, 1605

When I was very young I asked my father: 'What is the largest number you know?' and he answered 'octillion'. At the time I diid not know any compact notation for writing large numbers, such as writing them in powers of 10, but I soon decided that an octillion, whatever it might look like, must be too small to be the largest number. After all, if you add 1 to it you get a bigger one!

The obvious answer, I decided, was to count to infinity, or at least far enough to find a number for which I would need to invent a new name. I resolved to try this in bed that night, but sleep overcame me soon after the 12, 000th sheep. But I was now clear that there is no such thing as 'the largest whole number'.

And that is something that strikes me as quite profound. After all, our senses provide us with information that reflects the finiteness of our surroundings—even if they may seem forbiddingly large when you're six years old—yet here we have a system of allocating names, or symbols, that is essentially without any limits. How does this system of numbering, abstracted by us as a collective mental construct, reflect the finite physical environment? Or does our perception of the need for counting not originate in the observation of our immediate surroundings?

Counting numbers were used for practical purposes in pre-historic times. Whether hunting or gathering, farming or trading, even in battle, no-one could readily escape the need to distinguish between 'one', 'two' and 'many': for example, when describing a pack of wolves, the day's wild fruit pickings, a flock of sheep, sacks of corn offered for exchange, or the size of the enemy's clan. Simple tallying, such as recording the number of any group with notches on a stick or a collection of pebbles, probably preceded the actual naming of numbers, but at some stage the need to invent verbal descriptions of the counting process became unavoidable.

 https://doi.org/10.11647/OBP.0236.12

Today, we have become so used to our decimal system of naming the numbers we use every day that questions about the origin of their names seldom enter our consciousness—we learn them at mother's knee, at the same time as the alphabet. Usually we start by counting on the fingers of both hands.

1. Naming large numbers

In any event, I might have had trouble working out what 'octillion' could mean, since even now there is no universal agreement about the names of various (fairly) large numbers! Everyone agrees that we call a thousand times a thousand a million. In the decimal system we 'add a zero' whenever we multiply to 10 and we have a convenient shorthand notation, using, for example, $1000 = 10 \times 10 \times 10 = 10^3$, where the exponent 3 simply shows how many times we multiplied by 10. Similarly, we write one million as $10^3 \times 10^3 = 10^6$. After that, however, different naming conventions emerge.

If, like my father, you are German, or any continental European, you would (today) stick to the Latin origins of the terms we might use when multiplying a million times—which means that we add six zeroes each time:

a billion is a 'bi-million', which is obtained by multiplying a million (10^6)

by a million (10^6), so it becomes $10^6 \times 10^6 = 10^{12}$,

a trillion (tri-million) is a million billion ($10^6 \times 10^{12} = 10^{18}$),

a quadrillion becomes a million trillion ($10^6 \times 10^{18} = 10^{24}$),

and we continue via quintillion, sextillion, septillion—adding 6 to the exponent each time—to reach octillion as 10^{48}.

You could go on to nonillion (10^{54}) and decillion (10^{60}), although these terms reached the *Oxford English Dictionary* only relatively recently and the dictionary entry currently stops there. Not to be outdone, Wikipedia lists 100 such number names—each a million times the previous one—up to centillion (10^{600}).

This continental European number naming scale is now known as the *long scale*. It certainly has a long history. One of the earliest consistent accounts of number names generated in this fashion occurs in a 1484 article *Triparty en la science des nombres* by the French mathematician *Nicolas Chuquet* (c.1445-c.1488) who mentions number names very similar to the above, up to 'nonyllion', and continues: 'and so on with others as far as you wish to go'.

The American version of these number names is known as the *short scale* In this scale, having reached a million, we then create a new number name when we've reached a thousand times the previous one. In terms of powers

Figure 1. 10^{21} pengo banknote[1]

of 10, in the short scale we create new names whenever we add 3 to the exponent. In this scale a billion is therefore a thousand million ($10^3 \times 10^6 = 10^9$),
although the Europeans confuse matters by calling 10^9 a *milliard* instead!

Continuing up the short scale, we reach a trillion as a thousand (short-
scale) billion ($10^3 \times 10^9 = 10^{12}$), so a short-scale trillion is the same number as a long-scale billion. And so it goes on, confusing us all. Just for the
record: repeatedly multiplying by 10^3 means that the short-scale octillion is
a mere 10^{27}—while in the long scale, 10^{27} becomes a thousand quadrillion,
so perhaps it should really be a 'quadrilliard'?

Currently, the USA, UK and Canada—and with them most other Anglo-
phone as well as Arabic-speaking countries—use the short scale, while most
countries in Europe, plus most French, Spanish or Portuguese-speaking countries elsewhere, prefer the long scale. Brazil is a rather large South-American
exception to this rule, while many Asian countries, notably China, India,
Japan, Pakistan and Bangladesh, employ different number-naming systems
altogether.

Number names beyond quadrillion (in either scale) are used fairly rarely.
During periods of hyperinflation—in Germany in 1923, Hungary in 1946,
or more recently, Serbia or Zimbabwe—some bank notes with very high
denominations were used briefly. The highest was a Hungarian banknote
nominally worth 10^{21} pengo, which was printed but never issued. Hungary
uses the long scale and, since $10^{21} = 10^9 \times 10^{12}$, the nominal value of the
note was shown proudly as one milliard billion pengo, or *'egy milliard b.-
pengo'*. (Had they used the short scale they could have called it 'sextillion
pengo'.)

In any event, there is no real need to invent names for large numbers,
since what we call *scientific notation* solves problems of this kind at a stroke,
simply by use of the decimal point and powers of 10. So we can write
$1,250,000$ ('one-and-a-quarter million') as 1.25×10^6, for example. Scientific notation enables us to compare large numbers quite simply: the *order
of magnitude* is given by the power of 10 (the *exponent*) we need to use when

---

[1]https://commons.wikimedia.org/wiki/File:HUP_1000MB_1946_obverse.jpg

we describe the number in this fashion. Thus, the estimated age of the universe is given as 4.32×10^{17} seconds, the number of stars in the observable universe is around 7×10^{22}, the most massive black hole so far observed is said to weigh some 8×10^{40} kilogrammes (recall that a kilogramme is 10^3 grammes) and so on.

And it works just as well for very small numbers: we simply replace positive exponents by negative ones (that is, we divide, rather than multiply, by various powers of 10). In this notation, *Planck's constant*, the 'quantum of action' in quantum mechanics, is $6.62606957 \times 10^{-34}$ (the units are metre-squared kilogramme per second, since you ask), while an electron 'weighs' about 9.11×10^{-28} grammes. So the truth is that we needn't really worry about 'naming' large numbers at all!

2. Very large numbers

I raised my innocent question about large numbers nearly five decades before the advent of Google, so perhaps it was not altogether surprising that my father was unaware that in 1920 the nine-year old *Milton Sirotta* had already invented the name *googol* for a large number that his uncle, the mathematician *Edward Kasner*, had dreamt up. A googol can be written as a 1 followed by a hundred zeros—or, more compactly, as 10^{100}.

In [24] Kasner reported that they had then invented *'googolplex'* as a number with 1 followed by 'writing zeroes until you got tired', upon which Kasner decided to allocate this name to the number with a googol of zeros; in other words, $10^{10^{100}}$. Of course, we can go on and on. For example, Google will tell you that a *'googolplexian'* has been defined as $10^{10^{10^{100}}}$, which is written as a 1 followed by a googolplex of zeros!

Why anyone should care, I am not sure, but perhaps we can ask Google. After all, its name is a misspelling of 'googol' (apparently the mistake occurred in 1997 while searching for an available internet domain name for the new company) and it cheekily misspells its headquarters similarly as 'Googleplex'.

In the last half-century even larger numbers have been devised, some of which were put to good use in advanced areas of modern mathematics. We will not attempt to discuss them here, except to say that several of these numbers are too large even for our scientific notation. Take, for example, *Graham's number*, which was devised by the US mathematician *Ronald Graham* in 1977 and for some years was regarded as the largest yet defined explicitly. It is so large that the *observable universe* is much too small to contain any decimal (base 10) representation of it, even if each digit is made unimaginably small—for example if each digit occupies only a single *Planck volume*—which is 4.2217×10^{-105} cubic metres (or, if you prefer, 4.2217×10^{-78} cubic millimetres).

Several leading mathematicians, notably *Donald Knuth*, have devised sophisticated notational 'shorthand' methods to describe such huge numbers. This, however, is likely to remain a distinctly minority sport!

But modern number enthusiasts have an illustrious forerunner; one who was active more than 2000 years ago.

3. Archimedes' Sand-Reckoner

In the third century BCE the Ancient Greek mathematician *Archimedes* (287-212 BCE)—arguably the greatest of all time—illustrated the power of mathematical reasoning by calculating an upper bound for the number of grains of sand needed to fill the known universe.[2] His paper, now known as the *Sand-Reckoner*, was addressed to *Gelon* (also known as *Gelo II*), the 'tyrant' (regent) of his home town of Syracuse in Sicily. It is a careful and quite accessible exposition of his calculations. It has been called 'the first research-expository paper ever written' (see [45] for details).

The standard translation [19] begins with a bold claim:

There are some, king Gelon, who think that the number of the sand is infinite in multitude; and I mean by the sand not only that which exists about Syracuse and the rest of Sicily but also that which is found in every region whether inhabited or uninhabited. Again there are some who, without regarding it as infinite, yet think that no number has been named which is great enough to exceed its magnitude. And it is clear that they who hold this view, if they imagined a mass made up of sand in other respects as large as the mass of the earth filled up to a height equal to that of the highest mountains, would be many times further still from recognising that any number could be expressed which exceeded the multitude of the sand so taken. But I will try to show you by means of geometrical proofs, which you will be able to follow, that, of the numbers named by me and given in the work which I sent to Zeuxippus, some exceed not only the number of the mass of sand equal in magnitude to the Earth filled up in the way described, but also that of the mass equal in magnitude to the universe.

3.1. Greek numerals. First of all, Archimedes had to develop a new system of numerical notation. Like our decimal system, the Greek *alphanumeric* system in his day used 10 as its base, but instead of developing a system of number symbols, the Greeks simply assigned different letters of their alphabet to successive numbers. Three of the letters used, representing our numbers 6 ϝ, 90 ꓷ and 900 ꓵ are obsolete and no longer appear in the Greek alphabet.

Numbers below 1000 are represented by means of 27 symbols as follows:

---

[2]In keeping with modern practice we will use the letters BCE (Before Common Era) rather than BC (Before Christ). Dates referring to the Common Era (previously denoted by AD, e.g. AD 750) will be referred to without prefix or suffix, e.g. Carl Friedrich Gauss (1777-1855).

Figure 2. Archimedes by Domenico Fetti, 1620[3]

$\alpha, \beta, \gamma, \delta, \varepsilon, \digamma, \zeta, \eta, \theta$ denote what we call 'units', $1, 2, 3, 4, 5, 6, 7, 8, 9$;

$\iota, \kappa, \lambda, \mu, \nu, \xi, o, \pi, \llcorner$ denote our 'tens' $10, 20, 30, 40, 50, 60, 70, 80, 90$;

$\rho, \sigma, \tau, \upsilon, \phi, \chi, \psi, \omega, \lambda$ denote $100, 200, 300, 400, 500, 600, 700, 800, 900$.

A number such as 243 is then given in *additive* notation as $\sigma\mu\gamma$, indicating that we should add together the numbers $(200, 40$ and $3)$ that are denoted by these three symbols. Similarly, what we would write as 571 is depicted by $\phi o\alpha$. The sum of these two numbers (814) is then written as $\omega\iota\delta$. Although this procedure may suffice for writing down the *result* of a simple calculation (possibly performed with an abacus or similar mechanical device) the actual *process* of addition is not easily memorised. By way of contrast, in our *positional* decimal number system, in writing the number two-hundred-and-forty-three as 243, we perform an addition, not with the symbols $2, 4, 3$ by themselves, but $(2 \times 100) + (4 \times 10) + 3$. Our ten number symbols are all we need, since the *positions* of the digits $2, 4, 3$ tell us that we mean two hundreds, four tens and three units.

Moving on to larger numbers, in the Greek system 'thousands' were expressed by preceding the corresponding letter used for units by a mark to its left: for example, $'\theta$ for 9000, so that $9,258$ would become $'\theta\sigma\nu\eta$ or, alternatively, $\overline{'\theta\sigma\nu\eta}$. Here the line above the letters indicated that one is dealing with a number rather than a word. This gave them specific symbols that combined to produce numbers up to 9999. The next number, $10,000$, was denoted by M. They could now combine these symbols and express larger numbers by using multiples of M and writing the multiplication factor above the letter M—for ease of typing this is shown as a 'power' in the

---

[3]https://commons.wikimedia.org/wiki/File:Domenico-Fetti_Archimedes_1620.jpg

following example:

$$30,254 = \overline{M^\gamma \sigma \mu \delta}.$$

The Greek word for the symbol M was $\mu\nu\rho\iota\alpha\varsigma$, later translated as *myriad* in Latin, and I will adopt the latter term. For his task, however, Archimedes needed a system in which much larger numbers could be expressed concisely.

3.2. Archimedes' number system. In the *long scale* version of our decimal system, once numbers up to a million have been named, one does not need a new number name until a million million, that is, a (long-scale) billion. The system Archimedes developed followed a similar pattern: he called the numbers up to a myriad myriads 'first numbers' and proceeded to make the final number the *unit* of his second system of numbers. In other words, numbers from then on are counted using *multiples of* this nunber. Expressed in terms of powers of 10, Archimedes' *first numbers* are all the numbers up to 10^8 (one-hundred million): since $M = 10,000 = 10^4$, a myriad myriads is $10^4 \times 10^4 = 10^8$.

In order to make sense of his system, Archimedes used the fundamental rule for multiplying powers; in our terms this rule is that, for any numbers a, b,

$$10^a \times 10^b = 10^{a+b}.$$

Having made 10^8 the new unit, or, as he called it, the *'unit of the second numbers'*, he was now able to keep counting until he reached a myriad-myriad times this unit, i.e. $10^8 \times 10^8 = 10^{16}$. This number now became the *'unit of the third numbers'* and he counted multiples of 10^{16} as his 'third numbers' reaching what we would call 10^{24}, since we can count up to 10^8 of these units. Then 10^{24} becomes the *'unit of the fourth numbers'*, etc., and we can continue until we reach the 'myriad-myriadth' unit.

This provides a very large number, obtained by multiplying 10^8 by itself 10^8 times, so we would write it as

$$(10^8)^{10^8} = 10^8 \times 10^8 \times .... \times 10^8$$

where the product on the right has 10^8 entries. We would write it as $10^{8 \times 10^8}$; written out it is 1 followed by 800 million zeros. Not yet satisfied, Archimedes then called all the numbers he had just defined the *'numbers of the first period'*, and again made the last one, namely $(10^8)^{10^8}$, the *'unit of the second period'*. Defining a new period each time, he could now construct a myriad-myriad periods, the last number therefore being

$$[(10^8)^{10^8}]^{10^8} = 10^{8 \times 10^{16}}$$

We would write this number as a 1 followed by 8×10^{16} zeros. In terms of the long scale, the *number of zeros* required to write out this number is eighty-thousand billion (or 80 quadrillion in the short scale). But no-one lives long enough to write it down in full: a day has $24 \times 3600 = 86,400$ seconds, hence, if a million people each wrote down one 0 every second, this collective would still need over 2500 years to complete the task! Thus Archimedes' system certainly names some very large numbers—but would it suffice to count the number grains of sand required to fill the universe?

3.3. Astronomical models. To determine this, Archimedes needed to decide on the astronomical model on which he would base his calculations. The models prevailing in his time were *geocentric*, placing the Earth at the centre of the universe and modelling planetary motions though a complex system of concentric spheres, rotating about the Earth at differing angles of rotation. As for the size of the universe, he begins by reminding Gelon of prevailing opinion:

Now you are aware that 'universe' is the name given by most astronomers to the sphere whose centre is the centre of the earth and whose radius is equal to the straight line between the centre of the sun and the centre of the earth. This is the common account, as you have heard from astronomers.

It is something of a puzzle (see e.g [11]) why Archimedes seems to claim that most astronomers of his time took the Earth-Sun distance as the radius of the 'universe', since the philosopher *Aristotle* (384-322 BCE) had asserted confidently, in his *Meterologica*, that *'the distance of the stars from the earth is many times greater than the distance of the sun'*. This work will have been known to Archimedes and his scientific contemporaries. Perhaps, in addressing his paper to King Gelon, who was probably more familar with astrology than astronomy, Archimedes felt that he had to acknowledge the layman's perception before contradicting it convincingly.

But Archimedes then draws Gelon's attention to an earlier proposal by *Aristarchus* for a *heliocentric* model of planetary motion, in which the Earth and the five visible planets orbit the Sun. Sadly, the original is lost, and Archimedes' comments comprise most of what we know about this proposal:

But Aristarchus of Samos brought out a book consisting of some hypotheses, in which the premisses lead to the result that the universe is many times greater than that now so called. His hypotheses are that the fixed stars and the sun remain unmoved, that the earth revolves about the sun in the circumference of a circle, the sun lying in the middle of the orbit, and that the sphere of the fixed stars, situated about the same centre as the sun, is so great that the circle in which he supposes the earth to revolve bears such a proportion to the distance of the fixed stars as the centre of the sphere bears to its surface.

Archimedes continues:

Now it is easy to see that this is impossible; for, since the centre of the sphere has no magnitude, we cannot conceive it to bear any ratio whatever to the surface of the sphere. We must however take Aristarchus to mean this: since we conceive the earth to be, as it were, the centre of the universe, the ratio which the earth bears to what we describe as the 'universe' is the same as the ratio which the sphere containing the circle in which he supposes the earth to revolve bears to the sphere of the fixed stars. For he adapts the proofs of his results to a hypothesis of this kind, and in particular he appears to suppose the magnitude of the sphere in which he represents the earth as moving to be equal to what we call the 'universe.'

As Archimedes notes, Aristarchus' stated assumption would lead to the conclusion that the fixed stars are 'infinitely far away', since the centre of a sphere (a dimensionless 'point') cannot be compared with the surface of the sphere.[4] He recognises that Aristarchus' model would yield a much greater radius for the sphere of the fixed stars than what he said was commonly assumed for the geocentric 'universe', namely the distance between the Earth and the Sun.

Thus, in order to find an *upper bound* for the size of the universe while continuing to work within a geocentric model (*'since we conceive the earth to be, as it were, the centre of the universe'*), he interprets Aristarchus' statement by equating the ratio of the diameter of the Earth ($d(E)$) to that of the Earth's supposed orbit around the Sun ($d(ES)$), with the ratio of the latter to the diameter of the sphere of the fixed stars ($d(S)$). This provides the equation

$$\frac{d(E)}{d(ES)} = \frac{d(ES)}{d(S)}.$$

Since $d(ES)$ must be much greater than $d(E)$, the diameter of the universe is now taken to be $d(S)$, i.e. the diameter of the sphere (centred at the Earth) containing the fixed stars. The equality means that in order to estimate $d(S)$ he only needs estimates for the other two diameters. This proved to be a more manageable task, although the details would lead us too far afield— for a modern exposition, see [45]. Armed with estimates for $d(E)$ and $d(ES)$, Archimedes was able to conclude that the diameter of the universe cannot be greater than 10^{14} *stadia* (a stadium amounts to about 180 metres in our terms).

3.4. Grains of sand to fill the universe. Finally, Archimedes has to estimate the size of a grain of sand and compute how many grains would fill

---

[4]Heath [20], p. 309, comments: While it is clear that Archimedes' interpretation is not justified, it may be admitted that Aristarchus did not mean his statement to be taken as a mathematical fact. He clearly meant to assert no more than that the sphere of the fixed stars is incomparably greater than that containing the earth's orbit as a great circle ; and he was shrewd enough to see that this is necessary in order to reconcile the apparent immobility of the fixed stars with the motion of the earth. The actual expression used is similar to what was evidently a common form of words among astronomers to express the negligibility of the size of the earth in comparison with larger spheres.

a sphere of radius one stadium. He first estimates that 40 poppy seeds, laid side-by-side, would measure approximately one finger-breadth (the Greek *dactyl*) which is about 1.9cm. A cube of this length contains $40^3 = 64 \times 10^3$ seeds. He then claims (without explanation) that, in volume, a poppy seed equals about a myriad (10^4) grains of sand, and promptly rounds up the product ($64 \times 10^3 \times 10^4$) of these numbers to 10^9. Finally, he rounds up to a myriad (10^4) the number of finger-breadths in a stadium—which is only a slight over-estimate this time. This gives him an upper bound for (the order of magnitude of) the number of grains of sand filling a sphere of diameter one stadium: since volumes change as the cube of the diameter, he (over)-estimates this as $(10^4)^3 \times 10^9 = 10^{21}$. Therefore, he concludes, the universe, having a diameter no more than 10^{14} stadia, can be filled up by using no more than $(10^{14})^3 \times 10^{21} = 10^{63}$ grains of sand.

Now 10^{63} is certainly a pretty big number. Yet, as Archimedes points out, this number is easily accommodated well within the *first period* of his numbering scheme: it is expressed as a *thousand myriad units of the eighth order of numbers*, which we would, in turn, express as $10^7 \times 10^{56}$ in our modern notation.

Today we do not work in terms of grains of sand, but use *nucleons* (the fundamental particles with mass making up the nucleus of an atom) and our best estimate of the number of nucleons making up the observable universe is in the order of 10^{80}. This is known as *Eddington's number*. And, since a grain of sand contains about 10^{17} nucleons, Eddington's number has the same order of magnitude as the number of nucleons contained in Archimedes' 10^{63} grains of sand! It would be wise, however, not to read too much into this surprising coincidence, especially since Archimedes' objective was not to find accurate estimates, but simply to show how very large numbers could be identified within a coherent nomenclature.

4. A long history

Having taken for granted the notion of *counting*, we have so far encountered only whole numbers and decimal fractions. Nothing has yet been said about basic arithmetic. Rather than begin such a discussion with a pre-ordained set of rules, such as those learned in primary school, I will explore the gradual development of arithmetic in a historical context to illustrate how our concept of number was widened repeatedly in order to describe all the possible solutions of various mathematical problems. In the process we will encounter different notational and conceptual approaches to the writing and manipulation of numbers, mirrored in the evolution of the expression of practical problems in mathematical terms, first in verbal descriptions and later by means of *equations*.

One example of this process is the gradual development of awareness that allowing only solutions consisting of positive whole numbers is an unsustainable restriction. Today, of course, handling *negative* numbers causes us no difficulties, familiar as we are with temperatures below zero and negative bank balances! In earlier times, mathematicians in different parts of the world struggled to accept negative numbers as meaningful entities and to devise rules for manipulating them. It was only in the sixteenth century that some European mathematicians began to accept negative numbers as meaningful entities.[5]

Although, as we shall see, the Ancient Greeks did not regard *fractions* as numbers *per se*, their deep and highly influential researches into geometry extensively employed *ratios* of two (positive) whole numbers as a way of measuring the relative sizes of quantities such as lengths, areas or volumes. While philosophers argued whether such ratios should be regarded as numbers or not, their practical significance ensured that they were studied in detail by early mathematicians. Greek mathematics, with few exceptions, remained focused on geometry rather than arithmetic; other early traditions, in Egypt and especially in Babylon, developed effective arithmetical techniques to handle many specific practical problems involving ratios.

Despite the dominance of rigorous Greek geometry in the surviving ancient texts, fortunately preserved and further developed by Arab mathematicians between the eighth and eleventh centuries, aspects of all these different traditions can be found in the transmission of the 'wisdom of the ancients' to early modern Europe from the twelfth century onwards. European mathematicians of the Renaissance readily accepted that fractions can be treated as numbers which can be added or multiplied. They recognised that one can always express a ratio of two whole numbers in 'lowest terms' by cancelling common factors, so that, for example, $\frac{2}{4}, \frac{3}{6}$, etc., all represent the same relationship as $\frac{1}{2}$, and lead to the same *rational number*.

On the other hand, an air of mystery continued to surround the results of a geometric construction (and, later, the nature of certain solutions of an equation) where the quantity required could *not* be expressed precisely in terms of a ratio of two whole numbers. Today we still call such numbers, like $\sqrt{2}$ or π, *irrational*. Defining irrational numbers rigorously in arithmetical terms (rather than describing them negatively, as 'not rational', or by means of geometric constructions) posed a continuing theoretical challenge, although mathematicians throughout the ages found ingenious ways of approximating these mysterious quantities to a high degree of accuracy

---

[5]In this they were much slower than their counterparts elsewhere. In China, for example, negative numbers appeared in the *Nine Chapters on the Mathematical Art* (Han dynasty, some 2000 years ago). Rules for their manipulation—including with rods of different colours for positive and negative numbers—were in place by the third century. See Footnote 1, **Chapter 1.**)

by what we now call rational numbers. The question how we should define irrational numbers as members of a logically consistent *number system* on which to base arithmetic, was only tackled consistently in the latter half of the nineteenth century. It took until the 1870s for the (almost) universally accepted *real number system* to be cast in its modern form in a way that could underpin modern mathematics and its many applications.

It will also take us quite a while to get there in this book. I will start at the beginning by looking at some of our earliest reliable evidence concerning number systems.

CHAPTER 1

Arithmetic in Antiquity

The monuments of wit survive the monuments of power.

Sir Francis Bacon, *Essex's Device*, 1595

Summary

In this chapter the focus is on two ancient civilisations: Babylonian and Greek. Our evidence for the former comes from a large number of sun-dried clay tablets (found in modern-day Iraq) that were only deciphered less than a century ago. By contrast, the mathematics and philosophy developed in the Greek city states (notably Athens) and surrounding territories, well over 2000 years ago, have underpinned Western civilisation ever since the Renaissance. The content of the thirteen books of Euclid's famous *Elements of Geometry* dominated Western school mathematics well into the twentieth century, usually giving school pupils their first experience of mathematical *proofs*. It remains a beacon of mathematical achievement in antiquity.

In *Babylonian arithmetic*, on the other hand, we find the first truly *positional* number system, essentially equivalent to our decimal system, although its *base* was 60 rather then 10. Traces of this system remain in our the division of an hour into 60 minutes, each of which has 60 seconds, for example. We begin the chapter with a brief glimpse of the ways in which this *sexagesimal* number system was used in the area around the Tigris-Euphrates valley to solve a variety of practical problems, notably including quadratic equations.

Mathematical development in *Ancient Greece* is traced back to *Pythagoras of Samos* (c.570-c.495 BCE), who was both a philosopher and a mathematician. Very little survives of the work of the influential quasi-religious *Pythagorean* sect he founded, except in occasional accounts by later commentators, of whom *Plato* (c.428-c.348 BCE) and *Aristotle* (384-322 BCE) are perhaps the most reliable. This chapter explores the group's philosophical claim that *'All is Number'* and the arithmetical techniques that led them to remarkable insights, such as the famous *Pythagoras theorem*, but also into logical difficulties. Their influence on the later work of the Athenian school around Plato, much of it preserved in Euclid's *Elements,* can be seen the latter's Books VII-IX and in an exhaustive study of *incommensurables* in Book X.

 https://doi.org/10.11647/OBP.0236.01

Remaining with arithmetic, the chapter closes with a brief look at the (much later) *Arithmetika* of *Diophantus* (c.210-c.290).

1. Babylon: sexagesimals, quadratic equations

Historical research relies on written records as its primary source of evidence. For this reason I omit mention of tallying or counting with sticks that precedes the earliest written records. Written records from early civilisations in China or India used materials that were not easily preserved, so that direct evidence of their work is scarce.[1] The best-preserved records from early civilisations are found on Egyptian papyri and hieroglyphs and on Babylonian clay tablets.

Most Babylonian tablets stem from the *Old Babylonian* period (1830-1501 BCE), others from the *Seleucid* period of the last three or four centuries BCE. A considerable number of mathematical clay tablets has been discovered. Some contain various tables of numbers, others describe recipes for solving specific numerical problems. Many are thought to have been used in schools training *scribes* for Babylonian society, which was probably an elite profession, open to a select few.

The tablets were inscribed in *cuneiform* script with a wedge-shaped stylus as shown in Figure 3—the name derives from *cuneus*, the Latin term for 'wedge'—and dried in the sun. The extent of their mathematical sophistication only became clear when cuneiform script was fully deciphered in the 1930s, much of it by the Austrian-American mathematician *Otto Neugebauer* (1899-1990), [34]. Earlier historians of mathematics had paid more attention to Egyptian geometry and arithmetic, although its impact on later mathematical development is perhaps less significant. For this reason Egyptian mathematics will not be considered here.[2]

The Babylonian number system combined 60 as the number base together with symbols for tens and units. For digits up to nine, the number was marked by that number of vertical wedges, and the number of multiples of 10 was marked similarly by up to five horizontal (or tilted) wedges. This enabled them to display numbers $1, 2, ..., 59$. We call such a number system *sexagesimal*, just as we use the term *decimal* for our usual (base 10) numbers, or *binary* (also *dyadic*) when using the base 2 (as in modern computing). The reason for the Babylonians' choice of 60 is not known, but the fact that $60 =$

---

[1]An account of Chinese mathematics and astronomy can be found in Volume 3 of Joseph Needham's multi-volume work *Science and Civilization in China*. See also *Chinese Mathematics, A concise history* by Li Yan & Du Shiran, (translated by J.N. Crossley and A.W.-C. Lun), Oxford, Oxford Science Publications, 1987, and the article 'Chinese Mathematics', by Joseph Dauben, in the volume edited by V.J. Katz et al.: *The Mathematics of Egypt, Mesopotamia, China, India, and Islam: A Sourcebook*. Princeton, Princeton University Press, 2007.

[2]For Egyptian mathematics see (e.g.): A. Imhausen, *Mathematics in Ancient Egypt. A Contextual History*, Princeton, Princeton University Press, 2016.

Figure 3. 10329 in cuneiform script

$2 \times 2 \times 3 \times 5$ has more divisors (in fact, twelve: $1, 2, 3, 4, 5, 6, 10, 12, 20, 30, 60$) than $10 = 2 \times 5$ (which has only four: $1, 2, 5, 10$) may have been a factor in this choice.

The key observation, nearly 4000 years ago, was that, once symbols for $1, 2, 3, ..., 59$ had been decided upon (and executed with no more than five horizontal and nine vertical wedge strokes), *all* other (whole) numbers could be understood with these symbols. To write numbers outside the range 1 to 59, the Babylonians used a *positional* (or place-value) system, breaking up the numbers according to successive powers of 60 and separating these by a space, as in Figure 3, which shows the number

$$10329 = 2 \times (60)^2 + 52 \times (60)^1 + 9 \times (60)^0.$$

(The final term is simply 9, since $n^0 = 1$ for any n. This follows from the power law $n^a \times n^b = n^{a+b}$, using $b = -a$.)

The spaces between each group of wedges indicate the relative power of 60 that each group occupies. Here we have implicitly assumed that we are dealing with a whole number.

However, the positional system was also used to include *sexagesimal fractions*. For example, the numbers

$$2 + 52 \times (60)^{-1} + 9 \times (60)^{-2} = 2 + \frac{52}{60} + \frac{9}{3600}$$

$$2 \times 60 + 52 + 9 \times (60)^{-1} = 172 + \frac{9}{60}$$

would be written exactly as the number given in Figure 3. As no space was left at the end of a number, its absolute size often had to be inferred from the problem under discussion, although in some tablets the number would be followed by a word indicating what power was intended for the final group of wedges. More seriously, in the Old Babylonian tablets there is no symbol for 0 to indicate the absence of a power (as would be needed in $7209 = 2 \times (60)^2 + 9$, for example), although some texts appear to indicate this by leaving an extra internal space.

By the time of the second major set, dating from the *Seleucid period* (the last four centuries BCE) the second ambiguity had been removed. The occurrence of zero was now indicated by a space marked with two small oblique wedges, showing that that particular power of 60 is 'skipped'. To write down 7209, the scribe would now replace the central group of wedges in Figure

3 (denoting 52×60) by two oblique wedges, rather than simply omitting it without further indication. However, this practice appears only to have been used when zero occurs in a intermediate position, such as in this example. It was not used at the end of a number, so the absolute size of the number would continue to be deduced from the context of the particular problem.

Despite its peculiarities, the Babylonians could use their system to add, multiply, subtract and divide numbers in much the same way as we do with decimal notation, and to treat fractional parts of the numbers in exactly the same way as the integral parts. This was a major notational and conceptual advance.

It is convenient to use Neugebauer's notation to express the sexagesimal system in our decimal symbols. For example, the above number $2 \times 60 + 52 + 9 \times (60)^{-1} = 172.15$ is written by Neugebauer as $2, 52; 9$. The powers of 60 are separated by commas, where Babylonians would use spaces instead, and a semicolon separates the fractional from the integral part.

A large proportion of the cuneiform tables that have been found contain arithmetical tables, listing, in sexagesimal form, squares, cubes, reciprocals and even square and cube roots of numbers. They probably served in the ancient schools for scribes as the precursors of the books of logarithmic tables that were prevalent in our secondary schools until a few decades ago, before being replaced by electronic calculators and computers.[3]

The cuneiform tables were necessarily incomplete: they dealt only with *regular* sexagesimals, i.e. numbers that could be expressed simply in sexagesimal form. This was not possible for certain fractions, such as the reciprocal of 7, for example. In a tablet containing a typical table of reciprocals one usually finds two columns, and the two numbers in the same row always have 60 as their product. But immediately following the row listing the numbers 6 and 10 (the 'reciprocal' of 6 is $\frac{60}{6} = 10$) we find the numbers 8 and $7; 30$, which represents $\frac{60}{8} = 7\frac{1}{2}$, written to base 60 as $7 + \frac{30}{60}$. The row that would contain 7 and its reciprocal is simply omitted. The reason for this is clear: $\frac{60}{7}$ cannot be written as a *finite* sexagesimal – when using 'long division', as in $\frac{60}{8} = 7 + \frac{30}{60}$ (in Neugebauer's notation: $7; 30$), the ratio $\frac{60}{7}$ cannot be expressed as a sum of the form $\frac{a_1}{60} + \frac{a_2}{(60)^2} + ... + \frac{a_n}{(60)^n}$ for any *finite* sequence of numbers $(a_i)_{i \leq n}$ of the numbers $\{1, 2, ..., 59\}$, since all remainders are non-zero.

The same problem arises in our familiar decimal notation: at school we all meet infinite 'recurring' decimal expansions such as $\frac{1}{3} = 0.33333....$ and $\frac{1}{7} = 0.142857142857....$ Decimal notation (that is, dividing $1.000000...$ by 7) requires the second of these to begin with the finite sum

---

[3]The invention and role of *logarithms* will be discussed in **Chapter 3**.

$$\frac{1}{10} + \frac{4}{(10)^2} + \frac{2}{(10)^3} + \frac{8}{(10)^4} + \frac{5}{(10)^5} + \frac{7}{(10)^6}.$$

The numerators $1, 4, 2, 8, 5, 7$ of the six terms repeat indefinitely, the sum of these terms is multiplied by $\frac{1}{(10)^{6k}}$ for $k = 0, 1, 2, \ldots$, and the results are then summed. Thus the expression in decimal notation even of simple rational numbers often leads to summing an indefinite number of terms.[4] Notice that, in the sexagesimal system, $\frac{1}{7}$ provides the only 'irregular' reciprocal among numbers below 10, wheras in the decimal system the reciprocals of $3, 6, 7, 9$ are all 'irregular'!

In practice, and in modern computers, we handle this problem by using 'rational approximation': we terminate the expansion after a set number of decimal places, giving us an approximation that is sufficiently close for our purposes. The Babylonians used the same principle. Babylonian approximations of irregular sexagesimal reciprocals could easily be given with a high degree of accuracy, as would be needed for calculations with large numbers, for example in astronomy. The use of base 60 has the advantage that good accuracy can be achieved in relatively few steps: for example, an error of at most $(\frac{1}{60})^4 = \frac{1}{12,960,000}$ (achieved after four steps) is usually negligible in practice.

The tables of reciprocals allowed division to be carried out easily: taking $\frac{a}{b}$ as the product $a \times (\frac{1}{b})$ would allow the scribe to 'look up' the reciprocal of b in a table and multiply it by a, while interpreting the product in terms of the correct powers of 60. Such techniques are well suited to handle arithmetic with large numbers and can be applied very effectively in calculations resulting from astronomical or navigational observations.

Going beyond reciprocals, cuneiform tablets have been found showing that the Babylonians knew general methods for approximating square roots. A simple but effective method to estimate $\sqrt{a}$ is to guess a first approximation, say r_1. If its square exceeds a (we write this as $r_1^2 > a$), we see that as a second guess the ratio $\frac{a}{r_1}$ will be too small. The arithmetical average of these two guesses, $r_2 = \frac{1}{2}(r_1 + \frac{a}{r_1})$, provides a better estimate, but will again be too large, so that $r_2^2 > a$.[5] Now repeat this process, starting with r_2 in place of r_1, and continue in this fashion. One quickly obtains a good approximation to $\sqrt{a}$.

In the collection held at Yale University, USA, the tablet today known as *Yale7289*, which dates from between 1800 and 1600 BCE, displays the approximation of $\sqrt{2}$ by $1; 24, 51, 10$. This equals r_3 if one starts with the overestimate $r_1 = 1; 30$ (i.e. $r_1 = 1.5$ in decimal notation, which gives $r_1^2 = 2.25$). Approximating $\sqrt{2}$ by $r_3 = 1 + \frac{24}{60} + \frac{51}{(60)^2} + \frac{10}{(60)^3}$ (which we would write

---

[4]We return to this issue in **Chapter 7**.
[5]See *MM* for a simple proof of this claim.

$$\text{𒁹 𒐏𒐖 𒐐𒐏𒁹 𒌋}$$

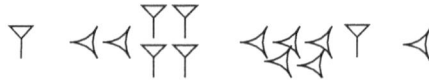

Figure 4. Approximating $\sqrt{2}$

as 1.4142162963 in decimal form) is accurate to 5 decimal places. The tablet is incomplete, and no workings are shown, but it seems plausible that the scribe might have used the above method.

While Egyptian papyri provide evidence that the solution of *linear* equations (i.e. of the form $ax - b = 0$, with solution $x = \frac{b}{a}$) formed a standard part of Egyptian mathematics, few Babylonian texts appear to deal with such problems. Instead, many tablets include more complex problems that lead to *quadratic* equations. A general solution procedure for quadratics is illustrated in several Old-Babylonians tablets, although they always deal with specific numerical problems. A text from the early Hammurabi period, for example, poses the problem of finding the side of a square, given that the area less the side is $14, 30$. This number is ambiguous, since we don't have a symbol for zero at this stage. We will read $14, 30$ as $(14 \times 60) + 30 = 870$, as the Babylonians often preferred to start calculations with whole numbers. If the side is x, the area of the square is x^2, and we must solve the equation

$$x^2 - x = 870.$$

Using Neugebauer's notation for numbers, we translate the scribe's instructions as: *Take half of* 1, *which is* 30, *and multiply it by* 30, *which is* 15. *Add this to* $14, 30$ *to get* $14, 30; 15$. *This is the square of* $29; 30$, *and the result is* 30, *the side of the square.*

To understand the quotation from the tablet, recall that we would write $\frac{1}{2}$ as $0; 30$ and $\frac{1}{4}$ as $0; 15$ instead of 30 and 15. The scribe probably found these from a table of reciprocals, given as the numbers whose products with 2, respectively 4, come to 60. To follow his procedure we reconstruct the general method in modern terminology. In the equation $x^2 - x = 870$ the coefficient of the term in x (the *linear* term) is -1, while 870 is the constant term. Call these b and c respectively, so that the equation we seek to solve is $x^2 + bx = c$. Following the scribe's instruction we now divide b by 2 and square the result, obtaining $(\frac{b}{2})^2$, which we add to c. We then take the square root (the words 'square' and 'square root' were used interchangably by the Babylonians) of this sum (probably looking it up in a table) and and subtract $\frac{b}{2}$ to find that

$$x = -\frac{b}{2} + \sqrt{(\frac{b}{2})^2 + c}.$$

This recipe can be derived from the following simple picture (although we have no direct textual evidence of any geometric figures that may have

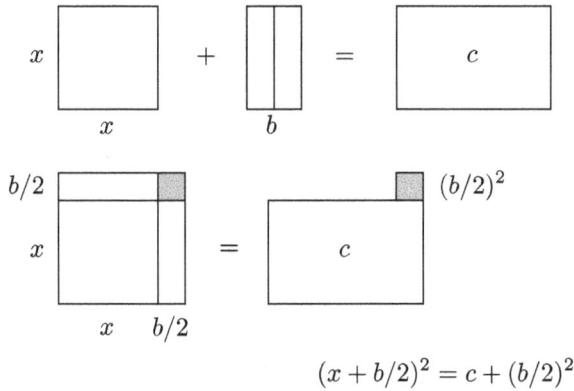

$$(x + b/2)^2 = c + (b/2)^2$$

Figure 5. Solving quadratic equations

been drawn): the equation $x^2 + bx = c$ says that by adding the square of side x to a rectangle with base b and height x, we obtain a given area c. To find x, we cut the base b of the rectangle in half, then arrange these two thinner rectangles on the square (one on top, one on the side, as in Figure 5, which is taken from [44]). We 'complete the square', which has the new base $x + \frac{b}{2}$, and to keep the two sides of the equation equal we need to add the small (black) square of side $(\frac{b}{2})$ to the area c. Taking the square root on both sides yields $x + \frac{b}{2} = \sqrt{(\frac{b}{2})^2 + c}$. The numbers used by the scribe are $b = -1$, $c = 870$. So $(\frac{b}{2})^2 + c = 870\frac{1}{4} = \frac{3481}{4} = (\frac{59}{2})^2 = 29\frac{1}{2}$, as claimed. Now, to find x, we subtract $\frac{b}{2}$ to obtain x as the solution of our quadratic equation.[6] Since $b = -1$, this means that we should add $\frac{1}{2}$ to $29\frac{1}{2}$ and thus obtain 30, as required.

It is important to emphasise that there is still much discussion amongst historians of mathematics on the proper interpretation of cuneiform tablets. The above discussion reflects one particular reconstruction. Nonetheless, it is clear that the tablets portray a society in which significant mathematical techniques were taught and used to solve relatively complex quantitative problems.

2. Pythagoras: all is number

Speculations about the origins of various systems for counting continue to occupy historians and philosophers today, and written evidence of such musings has also been preserved in Ancient Greek texts, though even these

---

[6]In *MM* it is shown how this procedure leads very simply to the general formula for the 'solution of the quadratic equation' we all learn at school.

are unlikely to have been the first to consider such questions. In the ancient text *Problems*, attributed to Aristotle, he ponders the reasons why in his time 10 seemed to be used 'universally' as the base for number names:

> *Why do all men, whether barbarians or Greeks, count up to ten, and not up to some other number, such as two, three, four or five, so that they do not go on to repeat one of these and say, for example, 'one-five', 'two-five', as they say 'one-ten'* [eleven], *'two-ten'* [twelve]? *Or why, again, do they not stop at some number beyond ten and then repeat from that point? For every number consists of the preceding number plus one or two, etc, which gives some different number; nevertheless ten has been fixed as the base and people count up to that.*[7]

He then lists some possible reasons that may provide insight into the familiar arithmetic of his time – which he attributes primarily to the *Pythagoreans*, followers of Pythagoras of Samos.

> *Is it because 10 is a perfect number, seeing as it comprises all kinds of number, even and odd, square and cube, linear and plane, prime and composite? Or is it because ten is the beginning of number, since ten is produced by adding one, two, three, and four? Or is it because the moving bodies are nine in number? ..... Or is it because all men had ten fingers....*

Aristotle's reference to *nine* 'moving bodies' could be an an allusion to the astronomical system developed by the Pythagorean *Philolaus* (ca. 470-385 BCE). This system was reported to postulate the existence of a 'central fire' around which the earth and the eight celestial bodies visible to the naked eye, namely the sun, moon, five planets and the 'sky' (the fixed stars), would rotate. The earth would revolve about the central fire daily, the moon monthly and the sun annually, thus explaining why sun and moon rise and set. In order to arrive at the number 10 – which had special significance for the Pythagoreans – Philolaus is said to have claimed the existence of a 'counter-earth', which he assumed to be situated directly opposite the Earth from the 'central fire', also revolving about it daily, and which therefore always remained invisible to us!

I now consider ideas attributed to the Pythagoreans, as reported by later commentators, a little further, not least to understand more about the 'kinds of number' Aristotle refers to. Greek mathematics, in its various guises, has been singularly influential in the development of the subject through the ages. Let us start with the origins of Pythagorean arithmetic.

2.1. Ratios and musical harmony. No first-hand written records of the discoveries of Pythagoras and his immediate followers survive today. Aristotle and his teacher Plato have a good deal to say – often highly critical and sometimes obscure – about Pythagorean beliefs and mathematical achievements. Their testimony on Pythagoras, though coming a good century after

---

[7]T.L. Heath, *Mathematics in Aristotle*, Taylor and Francis, e-book, 2011.

the fact, is distinctly more reliable than are the much later and highly partisan accounts produced by the so-called *neo-Pythagoreans*, who sought to resurrect and expand the elaborate *number mysticism* that Pythagoras' quasi-religious sect had initiated.

Our focus is on the *arithmetic* of the Pythagoreans, rather than on their mystical beliefs. Paradoxically, the major source for our understanding of the techniques of Pythagorean arithmetic is a work that does not deal primarily with arithmetic at all. It is the vastly influential treatise *The Elements of Geometry* (see e.g [21]), widely known simply as the *Elements* and produced in the Egyptian port city Alexandria by the mathematician *Euclid*.[8] The thirteen books of this work comprise the most widely studied mathematical text of all time, and were fundamental in shaping the subject throughout more than two millennia.

In Aristotle's *Metaphysics* we find a concise summary of Pythagoras' essential belief system:

in numbers, he thought that they perceived many analogies of things that exist and are produced, more than in fire, earth, or water: as, for instance, they thought that a certain condition of numbers was justice; another, soul and intellect, ... And moreover, seeing the conditions and ratios of what pertains to harmony to consist in numbers, since other things seemed in their entire nature to be formed in the likeness of numbers, and in all nature numbers are the first, they supposed the elements of numbers to be the elements of all things. (Arist. Met. i. 5.)

Here Aristotle refers to the speculations of *Empedocles*, who argued (ca. 450 BCE) that air, earth, fire and water made up the basic four elements from which everything was constructed. Aristotle refers to three of those, to contrast them with Pythagoras' view that *numbers* are the basic building blocks. Assigning numbers to various physical objects or concepts played a significant part in Pythagorean number mysticism.

Although detailed ancient references to Pythagorean arithmetic are not numerous, it is a widely held view that they concerned themselves extensively with *ratios*, which we will interpret in terms of ratios of positive whole numbers, i.e. positive fractions of quantities. Texts suggest that these explorations were prompted by empirical evidence that simple ratios of string or pipe lengths in musical instruments can produce harmonious sounds.[9] The Pythagoreans calculated that an octave must correspond to the ratio $2 : 1$, a fifth to $3 : 2$, a fourth to $4 : 3$ (we say 'two-to-one', three-to-two', etc.).

---

[8]We know very little about Euclid himself. The fifth-century commentator *Proclus* tells us that Euclid was active in Alexandria during the reign of *Ptolemy I Soter*, who ruled Egypt from 323 to 285 BCE. Euclid may have studied in Athens at Plato's *Academy*, and later established a substantial school in Alexandria. Most writers date the *Elements* as from around 300 BC.

[9]The most comprehensive translation of these ancient sources is found in the German text *Die Fragmente der Vorsokratiker* by H. Diels and W. Kranz (6th ed.), Weidmann, Dublin, 1952.

Their derivations, fortuitously preserved for us in various fragments
that appear as comments in another substantial work by Euclid, *The Division of the Canon* (usually known by its Latin name: *Sectio Canonis*), seem
to be based on two underlying postulates which they took as not requiring
further proof (see [38]):

(i) *musical intervals* [the differences in pitch between two notes] *can be
quantified by means of ratios of two (whole) numbers;*

(ii) *harmonic intervals* [intervals pleasing to the ear when two notes are
played together, such as in the above examples] *are characterised by ratios of
two forms: either $n : 1$ or $(n + 1) : n$, for some whole number n. Conversely, for
any whole number n, the ratio $n : 1$ produces a harmonic interval.*

The Pythagoreans had observed experimentally that octaves and double
octaves are harmonic, while repeated fifths and fourths are not, and also
that following a fifth by a fourth (or vice versa) produces an octave. With
the postulates (i),(ii), Pythagorean music theory can be derived quite simply,
using the *geometric mean* G of two given quantities a, b. This is defined via
the proportion $a : G :: G : b$ (in words: 'a is to G as G is to b'). We represent
this by the identity $\frac{a}{G} = \frac{G}{b}$, so that G is the solution of the equation $G^2 = ab$.

If the octave is given by the ratio $\frac{b}{a}$, the double octave $\frac{c}{a}$ must satisfy $\frac{c}{b} =
\frac{b}{a}$ since each ratio represents an octave. So b is the geometric mean of c and
a. But then $\frac{c}{a}$ *cannot* have the form $\frac{n+1}{n}$. Whenever three quantities a, b, c are
in geometric proportion, we have $c : b :: b : a$, so that with $c = n + 1, a = n$,
we would obtain $b^2 = n(n + 1)$ Since $n(n + 1)$ lies strictly between n^2 and
$(n+1)^2$, it cannot be a perfect square. So b cannot be a whole number. Hence
postulate (ii) ensures that the double octave $\frac{c}{a}$ has the form $m : 1$.

Next, consider the fifth and fourth. Both are harmonic intervals, so the
form of their ratios must be either $n : 1$ or $n + 1 : n$. If either of them had the
form $b : a = n : 1$, their compound ratio $c : a$ would be harmonic. But it was
observed empirically that double fourths and double fifths are *not* harmonic.
Hence the fifth and fourth must each have the form $(n + 1) : n$, (for different
$n > 1$) and their composition becomes the octave, as above. The simplest
numbers of the form $\frac{n+1}{n}$ are $\frac{3}{2}$ and $\frac{4}{3}$. Multiplying those provides $\frac{3}{2} \times \frac{4}{3} = \frac{2}{1}$,
so that the ratio 2 : 1 provides the octave.

The interval leading from the note a fourth up from the starting point to
the note a fifth up – described as a *whole tone* – became the principal unit in
the tonal scale. Since the ratios representing the fourth and fifth are multiplied when we add the intervals, subtraction of the intervals forces division
of the ratios, so that we obtain $(\frac{3}{2})/(\frac{4}{3}) = \frac{9}{8}$ as the ratio representing the
whole tone.

The earliest musical scale based on such simple numerical ratios is credited to Pythagoras himself, together with the discovery that the frequency
of a vibrating string is inversely proportional to its length. However, the

earliest reliable manuscripts on Pythagorean music theory stem from *Philolaus*, more than a century after Pythagoras. He derived the above ratios as well as the three intervals making up the fourth (or *tetrachord*), which are $9 : 8, 9 : 8, 256 : 243$. The adjustment to the final interval is required if one starts with two whole notes, to ensure that the final interval takes us to the ratio derived for the fourth: we need $\frac{9}{8} \times \frac{9}{8} \times \frac{a}{b} = \frac{4}{3}$, which leads to $\frac{a}{b} = \frac{8}{9} \times \frac{8}{9} \times \frac{4}{3} = \frac{2^7}{3^5} = \frac{256}{243}$. Such calculations led to what is today known as the Pythagorean *diatonic scale*, which Plato adpoted in constructing the 'world soul' in his *Timaeus*.

The symbolic notation we have used in this reconstruction was not used by the Pythagoreans. They and their successors did not perceive ratios as 'numbers' – this term was reserved for *multiples,* or what we would call the positive whole numbers $2, 3, 4, ...$ Euclid's *Elements* provide a strikingly vague definition of *ratio* as: *'a sort of relation in respect of size between two magnitudes of the same kind'.*

For the Pythagoreans, ratios were essentially a tool for *comparing* magnitudes, which were interpreted *geometrically*, as seen in Euclid's works. Numbers enter the discussion as multiples that tell us how often, in a ratio $A : B$, these quantities are 'measured' exactly by some (smaller) *unit*. Relating magnitudes A, B to a pair (m, n) of whole numbers (which we would regard as the fraction $\frac{m}{n}$) then signifies that the common unit 'goes exactly m times into' A and n times into B. Thus, in particular, for A and B to have a ratio, these two quantities must necessarily be of 'the same kind': both are musical intervals, or whole numbers, or geometric lengths, areas, or volumes, measured by a common unit. The unit itself is not regarded as a *number* in the same sense as the 'multiples' $2, 3, 4, ...$

However, the Pythagoreans could *compare* any two *ratios* $A : B$ and $C : D$, irrespective of whether these were 'of the same kind' (e.g. if A, B were lines, while C, D were areas). These four quantities are *in proportion* if the pair A, B, measured by some common unit (e.g. a length), generates the same pair of numerical multiples (m, n) as does the pair C, D, when measured by some other common unit (e.g. an area). This Pythagorean theory of proportions, largely preserved in Books VII to IX of Euclid's *Elements*, was central to their mathematical framework.

While the Pythagoreans discussed such arithmetical relationships verbally, without symbolic notation, they made considerable progress in their efforts to quantify musical relationships. Their triumphant conclusion was: *'All is Number'*. By this they meant that all natural phenomena can be understood in terms of the ratios of positive whole numbers. This turned out to be a rather sweeping conclusion, as they themselves discovered! Nonetheless, their ideas mark an important step (and the earliest that has been preserved) in humanity's attempts to describe natural phenomena systematically through *quantitative analysis*.

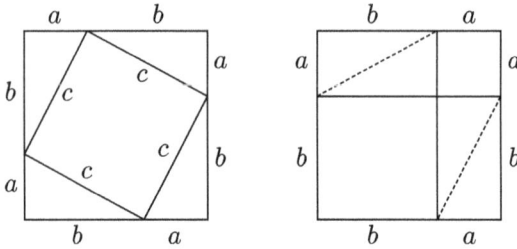

Figure 6. Pythagoras' Theorem

2.2. Pythagorean triples. For many, the name Pythagoras evokes memories of school mathematics, whether they be pleasant or painful! His famous theorem about the relationship of the sides of a right-angled triangle is probably the best-known result of Greek mathematics.

Pythagoras' Theorem

In any right-angled triangle the square on the hypotenuse is the sum of the squares on the other two sides.

If we denote the lengths of the sides by a, b, c with c as the hypotenuse, then this means that $a^2 + b^2 = c^2$.

A simple proof is illustrated in Figure 6, where we consider two ways of dividing up the square with side $(a + b)$. On the left, on each side mark off lengths a, b in order, starting at top right and going clockwise, and join points to produce four copies of the right-angled triangle with sides (a, b, c), situated around a quadrilateral whose sides all have length c, 'tilted' through the base angle θ between sides b and c of the triangle. At each vertex of this quadrilateral we have angles θ and $(90° - \theta)$ in the triangles that meet there, hence the remaining angle is a right angle, and therefore the tilted figure is a square.

On the right mark off the length a in both directions from the top right vertex, and similarly length b from the bottom left vertex, to construct squares with sides a, b, meeting in a point. What remains are two copies of the rectangle with sides a, b. The total area of the two rectangles equals that of the four right-angled triangles with sides (a, b, c) on the left, as they have the same base and height. Subtracting the triangles on the left leaves the square on the hypotenuse, while subtracting the two rectangles on the right leaves the squares on the legs of the triangle.

We have proved that $a^2 + b^2 = c^2$. (As Euclid would have put it: we have taken equals from equals, so the remaining areas are equal.) This proof has been called the 'Chinese proof' of the theorem, as it occurs in the ancient Chinese text Chou Pei Suan Ching. Euclid's *Elements*, Book I, has a quite different proof.

This relationship between the sides of a right-angled triangle was well-known to the Old Babylonians, who routinely made use of the theorem a thousand years before Pythagoras was born. Various tablets use it in a variety of problems: on a tablet now in the British Museum (BM85196), a beam of length 30 standing against a wall is said to have slipped from a vertical position so that the top has slipped 6 units. The scribe asks how far the lower end moved. Thus we have a right-angled triangle (a, b, c) with hypotenuse $c = 30$ and leg $b = 24$ units. To find a the scribe computes $\sqrt{(30)^2 - (24)^2} = 18$. The Babylonians applied this recipe, as one with general validity, in varied practical settings; the modern notion of verifying its validity diagrammatically stems from the later development of Greek geometry.

The best-known example displaying the depth of Babylonian understanding of the theorem is the tablet *Plimpton 322* in the Yale collection, which dates from 1800 BCE. Although now broken and incomplete, this lists a considerable array of triples (a, b, c) of whole numbers, now commonly known as *Pythagorean triples*, which satisfy the equation $a^2 + b^2 = c^2$. The simplest Pythagorean triples will be familiar: $(3, 4, 5)$ uses the smallest whole numbers possible, giving $3^2 + 4^2 = 5^2$ (our example above multiplies each side by 6). It is also easy to check that the triples $(5, 12, 13), (8, 15, 17)$ and $(7, 24, 25)$ are Pythagorean.

In Plimpton 322, the scribe's methodology in choosing his particular triples still leads to lively discussions among historians, but there is no doubt that he was familiar with very many such triples, including some with impressively large numbers, and that he arranged them in a consistent pattern, whose purpose we can only guess today.[10]

The two basic methods for generating Pythagorean triples were well known in Ancient Greece. In our terms they are:

(a) if $m > 1$ is an odd number, then $(m, \frac{1}{2}(m^2 - 1), \frac{1}{2}(m^2 + 1))$ is a Pythagorean triple;

(b) if m is an even number greater than 2, then $(m, (\frac{m}{2})^2 - 1, (\frac{m}{2})^2 + 1)$ is a Pythagorean triple.[11]

The influential fifth-century neo-Platonist commentator, *Proclus*, (while not notable as a reliable source, and writing nearly a millennium later) attributes (a) to Pythagoras himself and (b) to Plato. Both are easily checked

---

[10]See, for example, a trenchant rebuttal of earlier interpretations in Eleanor Robson: Neither Sherlock Holmes not Babylon: A Re-assessment of Plimpton 322, *Historia Mathematica* 28 (2001) 167-206. See [25] for an account of the tablet.

[11]For $m = 2$ the formula in (b) yields $(\frac{m}{2})^2 - 1 = 0$, which leads to the trivial triple $(2, 0, 2)$, corresponding to the 'triangle' with base angle 0. Note also that the triples do not all lead to distinct triangles. For $m = 3$ and $m = 4$ we obtain $(3, 4, 5)$ from the first formula and $(4, 3, 5)$ from the second.

by simple algebra: write out the squares in (a):

$$m^2+\frac{1}{4}(m^2-1)^2 = \frac{1}{4}[4m^2+m^4-2m^2+1] = \frac{1}{4}[m^4+2m^2+1] = [\frac{1}{2}(m^2+1)]^2.$$

proving that (a) is Pythagorean.

The proof of (b) is almost identical to the above and is left as a simple exercise for the reader. Readers allergic to algebra or who dislike powers higher than 3 (as did the Ancient Greeks) may safely skip these algebraic arguments. We will focus instead on simple geometric techniques by which the Pythagoreans may have derived these results.

2.3. Pebbles, triangles and squares. Four aspects of the arithmetic of the Pythagoreans are widely accepted as tools that were available to them. They

(i) used 'pebble arithmetic' for visual displays of number patterns,

(ii) regarded odd and even as *the two proper forms of number*,

(iii) used triangular, square and oblong numbers (for definitions see below),

(iv) explored Pythagorean triples.

An ancient (apparently Babylonian) technique of using an L-shaped figure, which the Pythagoreans called a *gnomon* (or stick and shadow), to generate particular number patterns also seems to play a significant part in their reasoning.

Books VII-IX of Euclid's *Elements* contain arithmetical results that are generally seen as exemplars of Pythagorean methods, although his proofs use geometrical figures contructed by straightedge and compass rather than diagrams consisting of groups of pebbles. (Here a *straightedge* is a ruler without marked lengths.)

Closely following [26], we now reconstruct some of these techniques in modern terms. Recall, however, that the visual character of early Greek mathematics (later expressed so elegantly in Euclid's *Elements*) meant that only *multiples* of the chosen 'unit', and *not the unit itself*, were regarded as actual numbers. In a geometric construction, a given length (area, volume) will *measure* an arbitrarily chosen *unit* (length, area, volume) a certain number of times.

Beginning with pebble arithmetic, one can combine (i) and (iii) above to define three kinds of *figurate numbers* reportedly used by the Pythagoreans to represent different numbers (see Figure 7). A whole number is said to be:

Triangular if it is represented in triangular form, using rows of $1, 2, 3, \ldots$ pebbles; $3, 6, 10, \ldots$ are examples.

Square if it is a perfect square, made up of equal numbers of pebbles in each row/column; e.g. $4, 9, 16, \ldots$

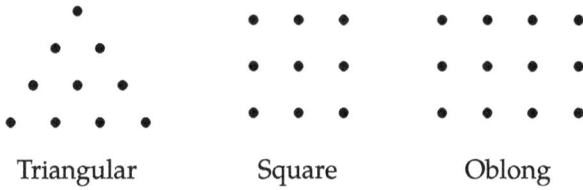

Triangular   Square   Oblong

Figure 7. Figurate Numbers

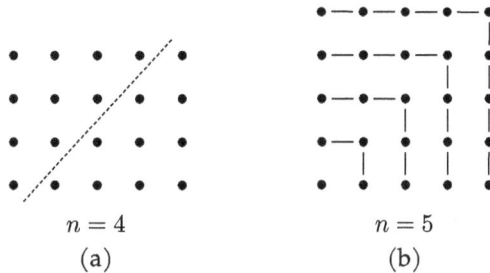

$n = 4$ $n = 5$

(a)   (b)

Figure 8. Sum formulae

Oblong if it is a rectangle, with one more pebble in one direction than the other, thus of the form $n(n+1)$; e.g. $6, 12, 20, ...$

Larger triangular numbers can be built simply by adding more rows - each row having one more pebble than the previous one. This immediately begs the question: how do we find the sum $1+2+3+...+n$? Does it help to look at our pebble representation? The triangular number itself appears not to provide immediate enlightenment. However, looking instead at the pebbles in an oblong number we can immediately find the answer—see Figure 8(a).

Drawing a *diagonal* (from just above the top right to just below the bottom left pebble) we have divided our oblong number into two equal pieces. But the area of the rectangle with sides n and $(n+1)$ units is obviously $n(n+1)$, as represented by our oblong number. Each of the two equal triangular pieces into which we have split this oblong number has n pebbles in its bottom row, so the nth triangular number is one-half of the oblong number $n(n+1)$. Summing over all the rows in the triangle, we obtain the familiar formula for the sum of the first n numbers:

$$1 + 2 + 3 + ... + n = \frac{1}{2}n(n+1).$$

Square numbers give us immediate insight into the sum of the first n *odd* numbers – see Figure 8(b). Successive square numbers can be built up

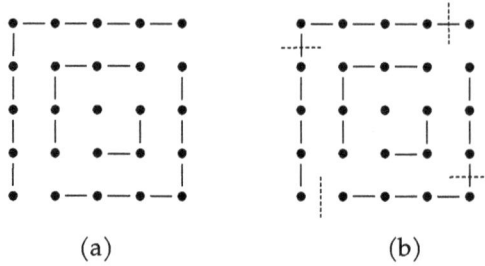

(a)   (b)

Figure 9. Putting gnomons about the unit

by starting with one pebble and adding successive gnomons – here taken as symmetrical L-shaped figures, the first having 3 pebbles, the next 5, then 7, etc. To obtain a square with n pebbles on each side we need the last gnomon to contain $(2n - 1)$ pebbles (the corner pebble serves both sides). In other words, we have demonstrated the identity

$$1 + 3 + 5 + ... + (2n - 1) = n^2.$$

These examples show that such summation formulae will have been well within the range of Pythagorean arithmetic.

2.4. Pebbles and gnomons. Aristotle discusses the Pythagorean practice of constructing square numbers by '*setting the gnomons round the unit*' - see Figure 9(a). In **[26]** this is interpreted as follows: start with one pebble at the centre, add a gnomon containing 3 pebbles below and to its right, follow this by 5 pebbles above and to its left, and continue alternating in this fashion, each time adding the next odd number of pebbles.

This idea leads one quite naturally to a 'pebble proof' that for odd M, the triple

$$(M, \frac{1}{2}(M^2 - 1), \frac{1}{2}(M^2 + 1))$$

is Pythagorean: since M is odd, so is its square M^2. Thus $M^2 \pm 1$ are both even, so all three the above are whole numbers. For any K, the difference of squares $(K + 1)^2 - K^2 = 2K + 1$ (which is also the number of pebbles in the gnomon we would add to the square of side K to obtain the next one). When M is odd we can therefore write as $M^2 = 2K + 1$. Now $K = \frac{1}{2}(M^2 - 1)$ is the side of the smaller square and $K + 1 = \frac{1}{2}(M^2 + 1)$ is the side of the larger square obtained by adding the gnomon M^2 to the smaller one. In other words, we have three squares, with sides, respectively, given by

$$(M, \frac{1}{2}(M^2 - 1), \frac{1}{2}(M^2 + 1)).$$

The sum of the areas of the first two squares equals that of the third, so these numbers form a Pythagorean triple.

For even M, similar ideas lead to the construction of Pythagorean triples of the form

$$(M, (\frac{M}{2})^2 - 1), (\frac{M}{2})^2 + 1).$$

For any even M, the square M^2 equals $4K$ for some K, and we can write this as $M^2 = (2K - 1) + (2K + 1)$. These are *two successive* gnomons, taking us from the square with side $(K - 1)$ to that with side $(K + 1)$. In other words,

$$(K + 1)^2 - (K - 1)^2 = 4K = M^2,$$

which again shows that the triple $(M, (\frac{M}{2})^2 - 1), (\frac{M}{2})^2 + 1)$ is Pythagorean. Figure 9(b) shows this for $K = 4$, with the perimeter of the larger square split into four equal parts.

There is no direct written evidence that such methods were actually employed by the Pythagoreans. The simple tools used here *suggest* that these results were within their range. The above arguments have an advantage over the purely algebraic proofs in that they do not involve powers greater then 2, rather than using fourth powers. Since early Greek mathematicians reasoned largely via geometric pictures, they had no use for powers beyond cubes, as there are three spatial dimensions.

Although other interpretations of the Pythagoreans' obsession with the number 10 have been given (see e.g. [44]), the triangle representing 10 may help us understand the quotation from Aristotle at the start of this section. The triangular number 10, the *tetractys*, (shown in Figure 7) has four rows, the top containing a single pebble (a *point*, dimension 0), the second two pebbles (two points define a *line*, dimension 1), the third three pebbles (three points define a *triangle* in the plane, dimension 2), and the bottom line has four pebbles (which define a *tetrahedron* in space—a triangular pyramid whose faces are four equilateral triangles, one serving as the base, with the other three meeting at the top vertex—dimension 3). The number 10, the sum of these four rows (the *tetrad*), thus represents the universe – while also serving as the unit for the *dekad*, the next higher order of counting (making 10 the base of the number system).

Figure 9(b) illustrates a simple result in what is today known as *number theory*: any square whose sides consist of an odd number of pebbles can be built from the unit (a single pebble) by adding pairs of consecutive gnomos around the unit. Each such pair can be split—as we did in Figure 9(b)—into 4 equal pieces. Hence an odd square number always leaves remainder 1 (the central pebble!) when it is divided by 4.

On the other hand, a square with even sides obviously divides into four equal pieces, each having sides with length one-half that of the original. So even squares are divisible by four.

Today we would describe these facts by saying that a perfect square leaves remainder 0 or 1 when divided by 4. This relationship is expressed as: $n^2 \equiv 0$ or 1 (modulo 4). In other words, when we divide 4 into a perfect square, we will *never* get 2 or 3 as the remainder.

This has consequences for right-angled triangles whose sides are whole numbers. It means that if C is even in a Pythagorean triple (A, B, C) then so are A and B : if both A and B are odd, the squares A^2, B^2 each leave remainder 1 upon division by 4, so the sum $A^2 + B^2$ leaves remainder 2, so the sum cannot equal C^2, which is divisible by 4. If exactly one of A, B is odd, then $A^2 + B^2$ would be odd, while C^2 is even. So if $A^2 + B^2 = C^2$ and C is even, we are forced to conclude that A and B are even.

So: if the hypotenuse C of a right-angled triangle with integer sides is even, then so are the other two sides. Next, suppose that *not* all sides (A, B, C) are even. Then C cannot be even, so it must be odd, by the above. If A, B were either *both* even or both odd, then $A^2 + B^2$ would be even, hence C would be even. So if any sides are odd, then C, and exactly one of A or B, must be odd.

We summarise this as a result that will be useful in Section 3 below – we will call it our

First Divisibility Lemma:

In a right-angled triangle whose sides (A, B, C) *are whole numbers:*

(i) if C is even, then are all three sides are even.

(ii) if any sides are odd, then C is odd, one of A, B is even and the other odd.

2.5. Side and diagonal. These 'pebble proofs' illustrate some of the techniques probably available to the Pythagoreans for the analysis of various geometric shapes as well as number relationships. However, Aristotle tells us that in this analysis they came across magnitudes that are incompatible with their bold claim that ratios of whole numbers (formed by multiples of a fixed unit), which were so useful in the analysis of musical harmony, could explain *all* natural phenomena. Since no original records remain to tell us how this came about, we again offer what constitutes one of several plausible scenarios for this discovery, rather than historical fact.

Construct a square of side 2 (in whatever units you prefer) and divide it into four unit squares by joining opposite midpoints of its sides. Divide each unit square into two isosceles right triangles by drawing diagonals that meet at the midpoints of the larger square. (See Figure 10).

Take any of the eight triangles. By Pythagoras' theorem, the square on its hypotenuse is $1^2 + 1^2 = 2$. Thus the hypotenuse, which is a line segment with length l, say) is the side of a square whose area is exactly double that

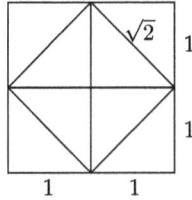

Figure 10. Side and diagonal

of a unit square.[12] So what is this length l–is it representable by a number, and if so, by what sort of number?

Aristotle answers this question with a logical argument showing that l *cannot* be expressed as a ratio of two whole numbers. It is his favourite example to illustrate *proof by contradiction*, an important proof technique we will come across repeatedly. He starts by assuming that the claim he is trying to prove is false, and shows that this must lead to a contradiction, so that the claim cannot be false, hence must be true.

To justify this logical principle, Aristotle argued that a proposition must *either be true or be false*. Thus either the proposition (P), or its negation $(not P)$, must hold. The assertion that these are the only possibilities is Aristotle's famous principle of the *'excluded middle'*. Like the great majority of modern mathematicians, I will side with him in this book, and utilise proof by contradiction frequently to justify my claims – although, as we will see in **Chapter 10**, this attitude is not quite universal.

Aristotle's proof that the diagonal of the unit square cannot be expressed as a ratio of two whole numbers goes as follows. Suppose that the relationship between the side and diagonal of the unit square *can* be expressed as a ratio of two whole numbers. (In modern terms, this amounts to the assertion that $l = \sqrt{2}$ is a rational number.) This would mean that we can write $l = \frac{a}{b}$ for some whole numbers a, b *with no common factors*.

Multiplying both sides by b, then squaring the results, we would have

$$a^2 = l^2 b^2 = 2b^2,$$

since the square on the side with length l has area 2. If a were odd, then a^2 would be odd, but in fact it equals $2b^2$. Hence a must be even and can be written as $a = 2c$ for some whole number c. Then $2b^2 = a^2 = (2c)^2 = 4c^2$,

---

[12]This fact is exploited famously in *Plato*'s dialogue *Meno*, where 'Socrates' teaches a young slave how to construct a square with double the area of a given one, while arguing that what the youngster was doing was not learning, but that he was simply 'remembering' a true statement that he had known subconsciously all along.

which, after cancellation, becomes

$$2c^2 = b^2.$$

This is turn shows that b^2, and therefore b, is also even.

So a and b have common factor 2, contrary to our assumption that they have no common factors. This contradiction shows that $l = \frac{a}{b}$ is impossible for whole numbers a, b.

But this proof, used by Aristotle as an example that would be well-known to his readers, seems rather unlikely as a *method of discovery*. In order to begin the above proof, one would need to suspect, at least, that l *cannot* be a ratio of two whole numbers. How might this possibility present itself to an unsuspecting Pythagorean?

Well, we have seen that the diagonal cuts the unit square into isosceles triangles, and our Pythagorean might wish to identify how such a triangle would produce a Pythagorean triple by adopting a suitable (smaller) unit length. The sides of the triangle would then be lengths expressible as whole number multiples (A, A, C) of this unit length, since the two sides meeting at the right-angle have equal length. Using part (i) of the First Divisibility Lemma proved at the end of Section 2.4, we see that, if the hypotenuse C is even then so is A, as in that case all three sides must be even. It that case we can halve each side and retain an isosceles triangle. Continuing to do this we must eventually arrive at an isosceles triangle whose hypotenuse C is odd. But now (ii) of the same result tells us that one of the legs of the triangle must be even and the other odd. But here the two legs have the same length A! This is a contradiction, proving that the ratio of diagonal to side in the unit square *cannot* be expressed as a ratio of whole numbers.

The fact that the side and diagonal of any square cannot simultaneously be multiples of the same unit means that they are not 'co-measurable' (or, more elegantly, *commensurable*) in terms of any chosen unit. They provide an example of two magnitudes (lengths) that are *incommensurable*. The Pythagoreans would presumably have found this highly disturbing, as they had no way of expressing the relationship between these two lengths by way of a number, yet it is clear that one can easily construct such lengths.

Of course, this is just one possible (if plausible) reconstruction of 'the discovery of incommensurables' by the Pythagoreans. There is a continuing debate among historians whether the ratio of diagonal and side of a square was actually the first quantity of this type to be considered. The fragmentary nature of surviving ancient texts and commentaries allows a variety of interpretations. Prominent among the alternatives put forward as the first known incommensurable quantity is the ratio of the diagonal to the side of a regular pentagon, often called the 'golden section' and given in modern terminology as $\frac{1}{2}(1 + \sqrt{5}) = 1.61803399...$ (see also **Chapter 2**, Section 2.2). I will not take sides in such disputes, but will continue to use $\sqrt{2}$ as

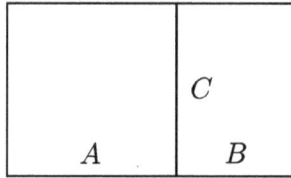

Figure 11. Sides and areas in proportion

our prototype of incommensurables, since it is the familiar example cited by Aristotle.

The existence of incommensurable magnitudes in geometry presents a problem for a Pythagorean world view that asserts that all objects can be measured in terms of ratios of whole numbers. As mentioned earlier, this world view is still visible, two centuries later, in the definition of *proportionality* that Euclid uses in Book VII of his *Elements*, where he reports and develops arithmetical results attributed to the Pythagoreans.

In Book VII, Euclid says that the magnitudes A, B and C, D are *in proportion* (recall that we denote this by $A : B :: C : D$) if A is '*the same multiple, part or parts*' of B as C is of D.[13] In other words, if $mA = nB$ for some natural numbers m, n, then we need $mC = nD$ for the two ratios to be in proportion. In our language, this says that D must be the same 'rational multiple' of C as B is of A: if $B = \frac{m}{n}A$ then $D = \frac{m}{n}C$, and we describe the ratio as the rational *number* $\frac{m}{n}$. For Euclid, on the other hand, ratio and proportion are not about 'numbers', so he does not express the ratio in this way.

With the above definition the following basic fact about rectangles can no longer be said to hold in general:

In a rectangle, a line parallel to one side divides the other side in the same proportion as the resulting areas.

This simply means that $A : B = AC : BC$ with A, B, C as in Figure 11.

This relationship between a ratio of line segments and the ratio of areas with these segments as their bases, while obviously true, can in some circumstances become a meaningless statement in the Pythagorean worldview: we need only take B as the chosen unit length and choose A as a length incommensurable with this unit. In modern notation, we might take $A = \sqrt{2} = C$ and $B = 1$. In that case the rectangles AC and BC comprise 2 and $\sqrt{2}$ area units respectively, while the line segments A, B have lengths of $\sqrt{2}$ and 1 linear units. But the claim that $\sqrt{2} : 1 = 2 : \sqrt{2}$ makes no sense

---

[13]A more general definition (due to *Eudoxus*) of proportion is given in Book V of the *Elements*, but it is generally accepted that in Book VII Euclid follows the Pythagorean concepts.

according to the above definition of proportions, since $\sqrt{2}$ is not 'a multiple, part or parts' of 1 (nor 2 of $\sqrt{2}$), precisely because $\sqrt{2} = \frac{m}{n}$ is impossible for whole numbers m, n.

In his treatise *Topics*, Aristotle highlights this difficulty clearly and suggests that a way out would be to *alter the definition* of proportionality so that it can encompass the above. We will see later how this problem was solved conclusively, reportedly by *Eudoxus* (408-355 BCE). The new definition is central to Book V of the *Elements*.

Despite the difficulties posed by incommensurables, it clear that the problem posed by $\sqrt{2}$ did not stop early Greek mathematics in its tracks, although it may have been instrumental in shifting its focus decisively from arithmetic to *geometry*, which appeared to be where a systematic study of these newly discovered magnitudes could be undertaken.

3. Incommensurables

Plato, whose Academy was, in its time, the most influential philosophical school in ancient Athens, frequently phrased his writings in the form of dialogues between 'Socrates' and other characters to drive home the main tenets of his philosophy. In one such dialogue the mathematician *Theaetetus* appears as a young man, relating how a Pythagorean lecture on incommensurables inspired him to make a significant breakthrough in identifying an unlimited number of examples of such magnitudes. The dialogue is less a historical account than a graphic lesson illustrating Plato's philosophical beliefs.

3.1. The Theodorus lesson. The passage describes, in the words of the youthful Theaetetus, his reaction to a lesson given around 400 BCE by the Pythagorean mathematician *Theodorus* of Cyrene, in which he demonstrated the incommensurability of the square roots of $3, 5, 6, \ldots$ 'up to 17' with the unit.

It is worth reading an excerpt to gain insight into its strongly visual description of mathematical statements – the translation is from [23]:

THEAETETUS: *Theodorus was writing out for us something about roots, such as the roots of three or five, showing that they are incommensurable by the unit: he selected other examples up to seventeen—there he stopped. Now as there are innumerable roots, the notion occurred to us of attempting to include them all under one name or class.*

SOCRATES: *And did you find such a class?*

THEAETETUS: *I think that we did; but I should like to have your opinion.*

SOCRATES: *Let me hear.*

THEAETETUS: *We divided all numbers into two classes: those which are made up of equal factors multiplying into one another, which we compared to square figures and called square or equilateral numbers;—that was one class.*

SOCRATES: *Very good.*

THEAETETUS: *The intermediate numbers, such as three and five, and every other number which is made up of unequal factors, either of a greater multiplied by a less, or of a less multiplied by a greater, and when regarded as a figure, is contained in unequal sides;—all these we compared to oblong figures, and called them oblong numbers.*

SOCRATES: *Capital; and what followed?*

THEAETETUS: *The lines, or sides, which have for their squares the equilateral plane numbers, were called by us lengths or magnitudes; and the lines which are the roots of (or whose squares are equal to) the oblong numbers, were called powers or roots; the reason of this latter name being, that they are commensurable with the former [i.e. with the so-called lengths or magnitudes] not in linear measurement, but in the value of the superficial content of their squares; and the same about solids.*

SOCRATES: *Excellent, my boys; I think that you fully justify the praises of Theodorus, and that he will not be found guilty of false witness.*

In summary, Theaetetus is here distinguishing between square numbers (what we today call 'perfect squares') and all other positive whole numbers (conveniently lumped together as 'oblong', since he is thinking only about areas of rectangles). He argues that in the former case, where the area is a perfect square, the *sides* (whose length is the square root of the area) are commensurable with the unit, hence may, in true Pythagorean fashion, be called 'magnitudes'. In the latter case, however, the side of a square whose area is an 'oblong' number (hence equals that of a non-square rectangle) is commensurable with the unit 'in square only' and so is not a Pythagorean magnitude – since it is not a 'multiple, part or parts' of the unit length.

As a simple example of this terminology, 3 is represented by a 3×1 rectangle; the side of a square with this area is $\sqrt{3}$. Although the area, 3 $(= (\sqrt{3})^2)$ is obviously commensurable with the unit, the side $\sqrt{3}$ is not, as (according to Theaetetus) Theodorus showed. So: $\sqrt{3}$ and the unit are 'commensurable in square only'. Euclid also adopts this terminology in Book X of his *Elements*.

In other words, the square root of a positive whole number either is itself a whole number (so the original number is a perfect square), or else it is irrational (cannot equal the ratio of two whole numbers). Confusingly for us, Theaetetus describes irrational square roots as 'roots' or 'powers' to distinguish them from his 'magnitudes', but the distinction between the two classes is clear nonetheless. The passage does not contain any indication of the proof of Theaetetus' bold claims.

We state the key result announced by Theaetetus rather more succinctly in modern terminology, as:

Theaetetus' Theorem

(i) *$\sqrt{n}$ is incommensurable with the unit unless n is a perfect square.*

(ii) *$\sqrt[3]{n}$ is incommensurable with the unit unless n is a perfect cube.*

The second claim is contained in Theaetetus' remarkably nonchalant – almost throw-away – remark *'and the same about solids'*. Not surprisingly, Socrates is impressed: *'Excellent, my boys.'*

Theaetetus' claims can be proved in a manner analogous to that given by Aristotle for $\sqrt{2}$. However, rather than simply relying on the fact that a whole number can be either odd or even, this proof (which we give in **Chapter 8**) crucially makes use of what is now called the *Fundamental Theorem of Arithmetic*, which we will discuss in **Chapter 7**.

We may wonder how Theodorus' case-by-case analysis might have employed the limited techniques we ascribed earlier to the Pythagoreans; in particular, whether this throws light on the reason he stopped at 17. This question has been debated extensively among historians. Our account provides a brief glimpse of the reconstructions in [**26**].

The only tools to be used are: pebble arithmetic with figurate numbers, the duality of odd and even, and Pythagorean triples.

Figure 12 displays three examples. We will consider (a) and (c) here. See *MM* for (b).

Theodorus is reported as dealing with each square root individually and he 'stopped at 17', which could suggest that he encountered some difficulty with this case. We will see below why this might have been so.

Rather than look at each case in turn, we can use the duality between odd and even, to consider the numbers $1, 2, 3, ..., 16$ in four groups according to their remainders when divided by $4 = 2^2$. In each group, the numbers can be handled similarly.

First, $4, 8, 12, 16$ are divisible by 4, with $4, 16$ as perfect squares, and $\sqrt{8} = 2\sqrt{2}$, $\sqrt{12} = 2\sqrt{3}$. So, dealing with $\sqrt{2}$ and $\sqrt{3}$ also deals with these cases.

Next, $1, 5, 9, 13, 17$ leave remainder 1. Of these, 1 and 9 are perfect squares. Among numbers *before* 17, this leaves only 5 and 13. We give the proof for $\sqrt{5}$ below.

Thirdly, $2, 6, 10, 14$ leave remainder 2 when divided by 4, so they have the form $4N + 2 = 2(2N + 1)$. Of course, $\sqrt{2}$ has already been done. Proofs for the remaining cases are slightly longer – for $\sqrt{6}$ see *MM*.

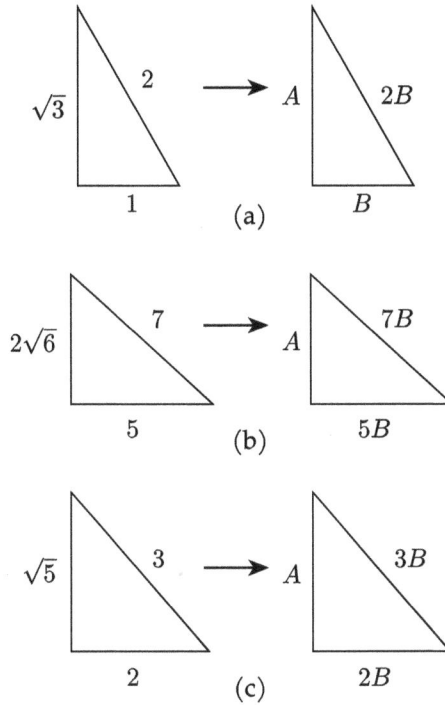

Figure 12. Theodorus' lesson

Finally, $3, 7, 11, 15$ leave remainder 3, so have the form $4N + 3$. Here we again just consider the first of these.

The 'Theodorus proof' given below for $\sqrt{5}$ (and in *MM* for $\sqrt{6}$) requires a lemma on divisibility by 4.

Second Divisibility Lemma:

In a Pythagorean triple (A, B, C) where not all numbers are divisible by 4, the only number divisible by 4 is either A or B.

This result follows readily from the duality of odd and even. A 'pebble proof' can be found in *MM*.

Thus $\sqrt{3}$ and $\sqrt{5}$ provide typical examples to illustrate what may have been Theodorus' approach.

Begin with $\sqrt{3}$. If there are multiples of the unit, A, B, such that $\sqrt{3} : 1 = A : B$, then the right-angled triangle with legs A, B gives rise to a Pythagorean triple $(A, B, 2B)$, since $(\sqrt{3})^2 + 1^2 = 2^2$, so that $A^2 + B^2 = (2B)^2$. The two triangles in Figure 12(a) are similar, as corresponding sides are in proportion. If A and B are both even, we can halve them successively until at least one of them is odd. But the hypotenuse $(2B)$ will then still be even, and, as in statement (i) of the First Divisibility Lemma (see the end of Section 2.4), this makes all three sides even. Since at least one side is odd, this contradiction shows that we cannot have $\sqrt{3} : 1 = A : B$, proving that (in our terms) $\sqrt{3}$ is not a rational number.

Next, we consider $\sqrt{5} : 1 = A : B$. As 5 is odd, our Pythagorean triple is $(\sqrt{5}, \frac{1}{2}(5 - 1), \frac{1}{2}(5 + 1)) = (\sqrt{5}, 2, 3)$, and our similar triangle has sides $(A, 2B, 3B)$, (see Figure 12(c)). Now if B is even, so is $3B$, which means that all sides are even, and the ratio $A : B$ is not in lowest terms. Hence we can assume that B is odd, in which case $3B$ is odd, so $2B$ is the only even side (by statement (ii) of the First Divisibility Lemma). This means that B is even, since by the Second Divisibility Lemma, the only even side, $2B$, must be divisible by 4. This contradicts our assumption that B is odd. Hence $\sqrt{5}$ cannot be rational.

Why does Theororus' method fail at 17? Observe that it is the first non-square number of the form $8N + 1$. If we attempt a 'Theodorus proof' that it is incommensurable with the unit, we would, as usual, assume that $A : B = \sqrt{17} : 1$. This will provide a triple $(A, 8B, 9B)$, since $17 = 9^2 - 8^2$.

But now our 'Theodorus' methods will no longer provide the desired contradiction: the hypotenuse, $9B$, is even only if B is even, and then all three sides will be even, so we can halve them repeatedly and confine ourselves to the case where B is odd. But then, with the odd hypotenuse $9B$, statement (ii) of the first Divisibility Lemma tells us that one leg is even and the other odd. This does not conflict with our triangle, since we now have A odd, $8B$ even and $9B$ odd. An appeal to the Second Divisibility Lemma will not help either: we know that not all sides are divisible by 4, as A is known to be odd. But now the other leg of the triangle, $8B$, is divisible by 4, while the hypotenuse, $9B$, is not. So the lemma cannot lead to a contradiction.

The two results we have relied on so far are not conclusive. We need a different analysis to deal with $\sqrt{17}$. But all we need is to consider divisibility by 3. So here is another divisibility lemma:

Pythagorean triples and divisibility by 3 :

In a right-angled triangle with sides (A, B, C), *if* 3 *divides the hypotenuse* C, *then it also divides* A *and* B.

For a 'pebble proof', consider a square number represented by pebbles. If its side is divisible by 3, then so is the square itself. Starting with a square of side $3n$, we build ever larger squares by adding gnomons: the first gnomon has $2(3n) + 1$ pebbles. The next gnomon is divisible by 3 (it has $2(3n + 1) + 1$ pebbles), so the resulting square also leaves a remainder 1 when divided by 3, while the third gnomon again produces a square with side divisible by 3. Thus a perfect square cannot leave remainder 2 when divided by 3.[14] Now suppose than $A^2 + B^2 = C^2$ and that 3 divides C. If neither of A, B is divisible by 3, then $A^2 + B^2$ leaves a remainder of 2, which is impossible, while if only one of them is divisible by 3, we cannot have $A^2 + B^2 = C^2$, since C^2 is divisible by 3. We have shown that 3 divides both A and B, which completes the proof of the lemma.

We apply this lemma to the triangle with sides $(A, 8B, 9B)$. The hypotenuse $9B$ is divisible by 3, hence so are A and $8B$. This means that B is also divisible by 3, hence A, B have 3 as a common factor and are not in lowest terms. This contradicts the choice of A and B, since we always assume them given in lowest terms. The contradiction shows that $\sqrt{17} : 1 = A : B$ is not possible for whole numbers A, B, so that $\sqrt{17}$ is irrational, as claimed.

The above arguments suggest that a first step towards the 'generalisation' needed for the proof of the result announced by Theaetetus is to consider divisibility by 3. This takes us past 17. To continue indefinitely, however, we must assume that the whole numbers represented by A, B have no common factors at all, rather than simply avoiding 2 as a common factor, as Theodorus does.

3.2. Euclid's classification. This excursion into early Greek arithmetic 'by reconstruction' illustrates how the number concept can be widened progressively through the analysis of particular problems—even if, as in this case, the mathematicians of the time reacted by refusing to regard the 'new magnitudes', the incommensurables, as numbers, and decided that they should be studied by geometric methods instead.

The later books of Euclid's *Elements* give a clear account of the degree to which they succeded in providing such a classification. The straightedge-and-compass constructions developed in the *Elements*, producing what we would today describe as various combinations of square roots, are extensive. Book X, by far the longest of the 13 books, is in large measure a compendium of these techniques. It includes 115 propositions, almost a quarter of the 465 propositions contained in the *Elements*. Since the constructions presented there are often quite complex, and have in any case been superseded by algebraic descriptions of the quantities in question, we will not attempt any proofs.

Euclid begins his definitions by distinguishing (just as Theaetetus had done in our extract from Plato) between *commensurable* magnitudes (multiples of the same unit) and magnitudes that are *commensurable in square only* (i.e. in our terms, the ratio $A : B$ does not equate to a rational number, but $A^2 : B^2 = m : n$ for some natural numbers, giving equal areas, $nA^2 = mB^2$).

We should recall that Euclid does not deal directly with numbers, but with geometric magnitudes. So, saying that two lengths are *commensurable*

---

[14]In modern notation (see also Section 2.4): $n^2 \equiv 0$ or 1 (modulo 3).

in square only means that they are incommensurable, but the squares on them 'are measured by the same area', as above. He starts by assuming a fixed *assigned line* (which serves as the unit of measurement) and calls a line *rhētos* if it is commensurable with the assigned line (either in length *or* in square only). He refers to an area commensurable with the square of the assigned line by the same term. Following [13], we use the term *expressible* to translate *rhētos*.

Expressible lines and areas (which are, at worst, square roots of rational numbers in our language) represent the 'easy case' for Euclid; his main interest is in classifying the *inexpressible* lines and areas. The first 18 propositions of Book X provide a detailed account of the properties of expressible magnitudes.

The next eight deal with the first subclass of inexpressible magnitudes, the medial: a *medial area* is equal to a rectangle with expressible sides commensurable in square only, and the side of a medial square is called a *medial line*. To express this in modern notation, fix the length of the assigned line as a. The lengths a and $\sqrt{2}a$ (the side and diagonal again!) are incommensurable, but commensurable in square, since the squares on these sides are a^2 and $2a^2$. The rectangle with sides a and $\sqrt{2}a$ has area $\sqrt{2}a^2$, so the square with this area has side $\sqrt[4]{2}a$. So the whole collection of *fourth* roots forms part of Euclid's class of medial lines.

He next considers sums and differences of incommensurable lengths and areas. In our notation this includes surds like $\sqrt{2} \pm \sqrt{3}$ or $1 \pm \sqrt{5}$. He shows that neither can be a medial. He calls the sum a *binomial*, the difference an *apotome* (terms still used in the sixteenth century by *Cardano* – see **Chapter 2**) and subdivides each of the binomials and apotomes into six different classes, analysing the relations between them and the medial.

In modern terminology what he deals with are sums and differences, repetitions and other combinations of various irrational square roots. Altogether, his classification amounts to the identification of 23 different classes of incommensurables, all of which can be recovered as solutions of polynomial equations with integer coefficients and of degree at most 4.

One motivation historians frequently cite for Euclid's extensive classification is that it includes, in particular, all the incommensurables needed for the construction of the five Platonic solids—the *tetrahedron, cube, octahedron, dodecahedron* and *icosahedron*—which Euclid achieves in the final books of the *Elements*. It turns out that comparing the edges of these three-dimensional figures, especially the last two, with the diameter of the sphere in which he assumes them to be inscribed, requires much of the sophisticated analysis Euclid develops in Book X.

4. Diophantus of Alexandria

While geometry remained the focus of most classical Greek mathematics, Archimedes, not all that long after Euclid, had a freer approach to the number concept, as testified in his *Sand-Reckoner* and in his approximations of particular irrationals, such as his remarkably accurate estimate that π lies strictly between $3\frac{10}{71}$ and $3\frac{10}{70}$.

More consistent moves away from the visual approach emerged during the Hellenistic period, from the death of Alexander the Great in 323 BCE to the battle of Actium in 31 BCE. This battle confirmed the dominance of Octavian – the future Emperor Augustus – over the forces of Mark Antony and Cleopatra. Early in this period the Egyptian port of Alexandria (established by Alexander the Great in 331 BCE) became a pre-eminent centre of learning. The city maintained this exalted status for nearly six centuries, throughout much of the Roman Empire. Major fires – the first during Julius Caesar's invasion of Egypt in 48 BCE – led to the destruction of the large Royal Library and the probable loss of the bulk of its estimated 400,000 manuscripts. Its daughter library, housed in the Serapeum (a pagan temple) survived and protected Alexandria's status as a centre of learning for another three centuries. It was finally destroyed in 391, following the Roman Empire's adoption of Christianity, under Emperor Constantine in 313.

Fortunately, significant parts (at least 6 out of a reported 13 volumes) of one of the most influential mathematical works written near the end of the period, the *Arithmetica* of *Diophantus* (ca. 210 to ca. 290) have survived. Four further books were found in a ninth-century Arabic transcription in 1968; these are thought by some scholars to include translations of later notes on Diophantus' work made by *Hypatia* (ca. 370-415), the first known female mathematician, who was killed by a Christian mob during conflicts in Alexandria in 415.

Despite its title, the *Arithmetica* is, in effect, a substantial work in what we might call *algebra* (although written more than five centuries before the term itself existed). Diophantus invented a system of scribal abbreviations (rather than a symbolic notation) to describe, analyse and solve various types of equation. He also broke with Greek tradition by considering powers higher than the third, removing the constraints imposed by geometric interpretations. He listed categories of numbers—which he still described as *made up of some multiple of units,* although, as we shall see, he accepted *ratios* of these among the solutions of various problems he posed. His notation for various species of number he particularly sought to investigate started with *squares* (designated by capital delta, Δ) and *cubes*. He went on to define *square-squares* (fourth powers), indicated by $\Delta^\Upsilon\Delta$ (two deltas together with a separating index) as well as higher powers up to *cube-cubes*. His notation distinguished carefully between what we would today call variables and constants, and he discussed problems involving various equations, expressed verbally and solved using his abbreviated notation. Diophantus consistently used symbols for the 'unknown' variable, but he usually assigned specific integer values to the constants he considered. He studied systems of linear as well as quadratic equations (which, as we have seen, have their origin in Babylonian times and also appear, in geometric guise, in Book II of Euclid's *Elements*), cubic equations and beyond, up to the sixth degree (*cube-cubes*).

He was only interested in positive rationals as solutions (there is no evidence that he accepted negative numbers), and frequently posed *indeterminate* problems, with more unknowns than equations, where the conditions imposed by the equations do not specify a unique solution. In complete contrast to Euclid's *Elements*, the *Arithmetica* contains no theorems, but consists of a series of solved problems that display various methods by which one may find two or more unknowns, usually under the requirement that certain expressions of them result in perfect squares or perfect cubes.

For example, in Book II, Problem 8 asks the reader: *To divide a given square number into two squares.*

In other words, we want to find (positive rational) x, y that satisfy $x^2 + y^2 = b^2$ for some given number b. Typical of his methods is that Diophantus chooses a specific example, $b = 4$, so that $x^2 + y^2 = 16$. His general method of solution seeks to ensure that either the quadratic term or the constant term in the equation disappears. He first notes that if the first square is x^2, the other is $y^2 = 16 - x^2$, and then says that y should be taken in the form $ax - b$ for some whole number a, again using $y = 2x - 4$ as his specific example. In this case he obtains

$$16 - x^2 = y^2 = (2x - 4)^2 = 4x^2 + 16 - 16x$$

and the constant term (16) cancels, so that, grouping like terms together,

$$5x^2 = 16x.$$

Therefore $x = \frac{16}{5}$ and $y = \frac{12}{5}$ would solve the problem, and the two required squares are $\frac{256}{25}$ and $\frac{144}{25}$, whose sum is $\frac{400}{25} = 16$.

This example shows that, while apparently unwilling to consider consider negative quantities as 'numbers', Diophantus nonetheless makes use of an assertion that is commonly learnt in primary school today. This is the familiar claim that '*a minus times a minus is a plus*'.

This is evident when he multiplies out $(2x - 4)^2$ and obtains the term $(-4)(-4) = 16$. Diophantus' notation differs from ours: he invents a 'subtraction sign', which looks like a capital Greek letter lambda (Λ) with a vertical line through the middle – it may be an abbreviation of *lepsis*, which means negation. He has no symbol for addition, so terms after the subtraction sign are simply grouped together. Nevertheless, without specific mention, he correctly applies the convention that multiplying out two terms starting with a negation sign must result in a term without one. This is fully in keeping with Aristotle's principle of the *excluded middle* (cf. Section 2.5): if the negation of a proposition is false, the proposition itself must be true; that is, *a proposition is the negation of its negation.*

As we can see, Diophantus was satisfied with obtaining *rational* numbers as solutions for the problems he posed. He thus moved beyond the restrictions imposed by classical Greek mathematicians, in effect regarding rational numbers as numbers in their own right, for instance by accepting rationals such as $\frac{256}{25}$ and $\frac{144}{25}$ as 'squares', which provide the solution of the above problem II.8. In similar fashion, various rationals of the form $\frac{m^3}{n^3}$ would be included included among his 'cubes'. However, he never articulated a fully consistent system of *adding* or *multiplying* fractions; nor did he use fractions explicitly when formulating his problems, or in arithmetical operations to simplify the equations he set up to solve these problems.

In modern number theory, Diophantus' name is invoked today for the modern field of *diophantine analysis*, which was mostly inspired by the work of *Pierre de Fermat* (1601-1665). In this field of research attention is focused on the more difficult task of finding *integer* solutions to indeterminate problems.

CHAPTER 2

Writing and Solving Equations

..to invent is to discover that we know not, and not to recover or resummon that which we already know.

Sir Francis Bacon, *The Advancement of Learning*, 1605

Summary

We review the development and acceptance of our current decimal system of number symbols, known to historians as *Hindu-Arabic* numerals. This reflects its Indian origins as well as its further development in the Arab caliphates that conquered the Middle East, parts of Central Asia, North Africa and Spain in the eighth and ninth centuries. In addition, while also taking initial steps towards *algebra* in a systematic study of quadratic and cubic equations, Arab scholars were crucial in the preservation and translation of Greek manuscripts.

The Hindu-Arabic numerals (together with most classical Greek mathematics) remained largely unknown in Europe until the twelfth century. A key figure in its transmission was Leonardo of Pisa (or *Fibonacci*), whose influence led to the replacement of reliance on the abacus by the use of Arabic numerals, with intermediate steps recorded on paper. This practice spread quickly, first in commerce, where symbolic notation served as shorthand. This, together with the study of equations, led to 'formulae' for the solution of cubics and quartics, first published in 1545 in the influential *Ars Magna* by Girolamo Cardano.

1. The Hindu-Arabic number system

While classical Greek texts were highly influential in cementing the dominance of geometry as the principal domain of certainty and proof, quite different sources influenced the development of arithmetical techniques and the symbolic representation of numbers. One key example was the early development, primarily in India, of the *decimal number system* we all take for granted today. However, piecing together this history today is complicated by an almost total lack of primary written sources, so that much of what is known is based on secondary sources.

 https://doi.org/10.11647/OBP.0236.02

A brief summary of results from recent scholarship on this topicis given in [25].[1] The classic text [31] describes in greater detail the gradual evolution, from origins in the Brahmin period in India (third century BCE) to early modern Europe, of the number symbols we use today. The earliest appearance of the nine number symbols that preceded the number digits $1, 2, ..., 9$ of our decimal system is found in decrees inscribed on pillars during the reign of King Ashoka (third century BCE). Over the next millennium the symbols were gradually transformed – together with the addition of zero by a dot (denoting 'absence') – although number symbols for $10, 20, ..., 90$ also remained in use, while combinations of symbols were initially needed to depict higher numbers. The earliest known Indian mathematician, $\overline{A}ryabhata$ (b.476) had a system of names for powers of ten, for example. The individual symbols for the nine digits were further developed by Arab mathematicians from the eighth century onwards, introduced to Spain during its Muslim occupation and transmitted via Italy to the rest of mediaeval Europe.

The crucial advantages of a place-value system using only nine symbols were realised quite early on. From at least 2000 years ago, columns on Chinese counting boards represented different powers of 10. It has been speculated that trading between the Chinese and Hindu cultures in South-East Asia may have led to an exchange of ideas, culminating (probably around 600) in the 'Hindu' numerals for numbers beyond 9 being dropped in favour of a full place-value system, including using the nine digits, together with a dot (later a circle) for zero. Evidence for this suggestion includes an inscription found in Cambodia dated to the year 683, shown as the 605th year of the Saka period and displaying the symbols then used for 6 and 5, separated by a dot. Similarly, an early eighth-century Chinese astronomical work explicitly describes the 'Hindu' use of a place-value system, including use of the dot, and comments that it made calculation 'easy'.

The earliest extant fragment referring to this 'Hindu system' comes from a Syrian priest, *Severus Sebokht*, who commented as early as the year 662 that the Hindus had a valuable calculation method 'done with nine signs'. He did not refer to the dot.

In any event, the most significant impact of Indian mathematics on modern mathematical techniques really began with the transmission of these numerals to the Arab world, apparently dating from about 770, when an Indian scholar visiting Baghdad showed his hosts a Sanskrit text containing calculation methods based on a decimal place system using nine number symbols and a symbol indicating zero (or 'absence'). Their use in the development of arithmetical techniques, including square root extraction, clearly impressed Islamic scientists, although it took a considerable time for the new methods

---

[1]This nomenclature 'Hindu-Arabic number system' has traditionally been used to describe aspects of mathematical development and innovation originating in the Middle East and/or the Indian subcontinent. Current debates on more appropriate terminology continue.

to gain wide acceptance in practice. The Sanskrit text was translated into Arabic, providing the basis of what gradually became widely known and eventually prized as *'Hisāb al-Hind'* ('Indian calculation').

As argued in [25], it seems likely that, while Indian mathematicians worked with a well-developed decimal place-value system for integers by the early eighth century, first steps in the extension of the system to include decimal fractions were due to somewhat later Arab mathematicians (perhaps beginning with al-Uqlidisi in the tenth century). Over the next few centuries Arab mathematicians also gradually modified the numerals, bequeathing to Europe what became the familiar number symbols of today's decimal system.

More generally, in the eighth and ninth centuries, a wide variety of scientific and mathematical texts was gathered together in Baghdad during the reign (786-809) of the caliph *Harun al-Rashid* and translated into Arabic. The different techniques and notations arising from these disparate sources were gradually harmonised and developed further. His influential *House of Wisdom*, became a public academy under his son, *al-Ma'mun* (reigned 813-33), who brought together Islamic as well as other scholars, often from Jewish, Zoroastrian or Christian backgrounds.

Mathematical research flourished in an atmosphere of free intellectual enquiry not dissimilar to that of the Greek city states. Although these researches were undertaken under a monotheistic regime, for at least two centuries secular enquiry was not regarded as creating conflict with religious dogma, and was openly encouraged. Nonetheless, such openness to new ideas was never fully universal throughout the caliphate and sadly it did not last. By the eleventh century the atmosphere had changed drastically. Greater religious orthodoxy in the *madrasas* focused more and more exclusively on Islamic law, to the effective exclusion of 'foreign sciences'. From that time onwards, mathematical activity other than basic 'practical' arithmetic, was increasingly discouraged, although some important exceptions remained, as described below.

Fortunately, by this time Europe was beginning to emerge from a prolonged period of turmoil and became receptive to the knowledge being transmitted, at first piecemeal but increasingly effectively, by scholars visiting the Arab world. European scholars discovered the work of Arab mathematicans such as *al-Khwarizmi* (ca. 780-850 AD) and his successors, who had adopted and gradually modified the numerals from India into a superior symbolic number system. The English word *algorithm* stems from a Latin transcription (*Algoritmi*) of his name. His most influential text deals with the simplification of equations, and our word *algebra* is a corruption of *al-jabr*, which describes the practice of transposing a term from one side of an equation to the other (with the attendant change of sign).

The earliest extant Arabic text (by *al-Uqlidisi*) shows that, by 962 AD, this number system included the symbol 0 and had a clear place value notation, as well as taking (still somewhat incomplete) steps towards handling decimal fractions. The author extols the flexibility and convenience of the 'Hindu' notation, and points to its advances over earlier 'finger-bending' methods to denote numerals, as well as to the ease of recording and checking complex calculations in the new system.

The final steps towards a fuller understanding of the decimal system, and of the convenience afforded by using the decimal notation when calculating with fractions, developed relatively slowly. However, treatises discussing approximations to fractions whose denominators include numbers other than 2 or 5 (so that the decimal expansion does not terminate), appeared in the Arab world in the twelfth century, showing that the authors clearly understood that the approximation can be made arbitrarily close by continuing far enough, providing *'an infinite number of answers, each being more precise and closer to the truth than the preceding one'*. The full decimal notation, with a vertical line separating the integral and fractional parts, first appears around 1500, in the work of *al-Kashi*. By then, Europe was catching up fast, with similarly sophisticated use of both finite and infinite decimal expansions, starting with the work of *Francois Viète* and *Simon Stevin* in the 1570s. All these developments played a fundamental role in creating the decimal number system we have today.

1.1. Arabic algebra. The great merit of the contributions of the Islamic mathematicians was not only that they preserved and transmitted the classical Greek works. They combined systematic geometrical methods developed by Euclid and his successors with earlier *ad hoc* methods inherited from the ancient Babylonians on linear and quadratic equations. In this, they constructed algebraic methods by which equations could be solved, yielding (as al-Khwarizmi already observes) *numbers* as the result of their calculations. Geometric arguments then served to *justify* the methods used—something which appears to be absent from Babylonian 'algebra'.

As noted above, the term 'algebra' stems from the term *al-jabr*, popularised in the famous work of al-Khwarizmi, *'Al-Kitāb al-muhtasar fī hisāb al-jabr wa'l muqabāla* ('The Compendious Book on the Calculation of *al-jabr* and *muqabāla*'), written around 825.[2] In his writings al-Khwarizmi does not make explicit use of equations, but describes the terms and steps in his solutions verbally – he was apparently unaware of the work of Diophantus.

Key to his solution of problems leading to quadratic equations of the form

$$ax^2 + bx + c = 0$$

---

[2]More details can be found in [6].

was the classification of problems involving an unknown, its square and constants (what he would call 'squares, roots and constants') into six basic categories as follows (the modern symbolic notation is given in brackets):

squares equal roots	$(ax^2 = bx)$
squares equal numbers	$(ax^2 = c)$
roots equal numbers	$(bx = c)$
squares and roots equal numbers	$(ax^2 + bx = c)$
squares and numbers equal roots	$(ax^2 + c = bx)$
roots and numbers equal squares	$(bx + c = ax^2)$

Note that this enables him to avoid negative coefficients, as well avoiding 0 on the right. His justification for the solution techniques he uses is always given in geometric constructions – essentially the 'completion of the square' that we discussed when looking at Babylonian cuneiform problems. For these, he compares two geometric figures, so that all quantities concerned must remain strictly positive.

A solution recipe (or 'algorithm') was then given for each of the six types. For the fourth (squares plus roots equal numbers) he gives the following example (here taken from [6]), where $a = 1, b = 10, c = 39$, giving the equation $x^2 + 10x = 39$:

You halve the number of roots, which, in this problem, yields five, you multiply it by itself, result is twenty-five; you add it to thirty-nine; the result is sixty-four; you take the [square] root, that is eight, from which you subtract half of the root, which is five, The remainder is three, that is the root of the square you want, and the square is nine.

In our notation his general solution of $x^2 + bx = c$ is therefore

$$\left(\sqrt{(\frac{b}{2})^2 + c} - \frac{b}{2}\right),$$

which is exactly what we saw in our earlier Babylonian example. Unlike the Babylonians, however, al-Khwarizmi takes great care to produce a geometric proof – in fact, he produces two proofs, one of which is virtually identical to our square-completion in Figure 5.

The influence of the 'six problems of al-Khwarizmi' is clear from subsequent texts throughout the Islamic period and early European mathematics – his solution methods were learnt by rote, and apparently more complex problems were systematically reduced (both by al_Khwarizmi and his successors) to one of the six types. He compared various combinations of 'squares, roots and numbers' and modified them by means of *al-jabr* (typically by moving terms to be subtracted to the 'other side', where they would be added), or *muqabāla* ('compensating'), which was done by reducing an apparently more complex equation by grouping like terms on the side where the net result is positive. For example, $5 + x^2 = 3x + 12$ is simplified to $x^2 = 3x + 7$. He also simplified both sides by cancelling common factors, finally arriving at one of the six types. Throughout, his comparisons were of aggregates of quantities that could be represented geometrically by means of squares and rectangles, so that the *dimensions* of the figures represented on each side remained the same.

From these beginnings, various generations of Islamic mathematicians fashioned systematic methods which they could justify geometrically in the Euclidean manner. After the translation of Diophantus' *Arithmetica*, late in the ninth century, their work went beyond the quadratic, to solve certain types of problems that we would see as leading to cubic equations and beyond. Although geometry remained the principal means of justifying their increasingly complex techniques and results, by the time of *Omar Khayyam* (1048-1131), Islamic mathematicians had studied and classified more than a dozen types of problems that we would today describe by means of cubic equations. In addition, in their studies of the salient features and differing advantages of the Babylonian sexagesimal number system (used throughout astronomy and astrology) and the new 'Hindu' decimal system, they made advances in unifying techniques for the manipulation of whole numbers and fractions, leading, in practice, to greater freedom in handling them both as 'numbers', even if they never fully articulated a consistent conceptual framework for these techniques.

Their ideas and techniques were to be taken up and ardently pursued by Renaissance mathematicians in Europe.

2. Reception in mediaeval Europe

In Europe, the fall of the Roman Empire in the fifth century resulted in its replacement by local feudal systems, led by often barely literate barons, who carved out local fiefdoms and engaged enthusiastically in local military campaigns. Latin survived primarily in Italy and what is now Southern France, where the Roman notion of the *quadrivium*—a term apparently coined by the Roman scholar *Boethius* (ca. 480-524) to describe the study of arithmetic, geometry, music and astronomy—was still regarded as necessary for the educated man, if only in a residual form with much-attenuated content.

During the reign of *Charlemagne* (742-814) in Central Europe the newly established multi-ethnic *Holy Roman Empire* began to develop a new focus on learning, based largely in the monasteries, under the leadership of *Alcuin of York* (735-804). He combined the quadrivium and the *trivium* (grammar, rhetoric and logic) into a comprehensive curriculum. The mathematical manuscripts available to Alcuin were few: the principal mathematical text circulating widely in Europe at this time was Boethius' *De Institutione Arithmetica*, a less than perfect version of an introductory text by the first century neo-Pythagorean *Nichomachus*.

The classical Greek works were to remain unknown in Western Europe until the twelfth century. Moreover, by the end of the ninth century, the brief revival of learning under Charlemagne had itself become overwhelmed amidst internal strife and by various invasions, from the East by Magyars, from the North by Vikings and from the South by Saracens. However, many of the monastic schools established by Alcuin survived these onslaughts and a more permanent revival began at the turn of the millennium, with *Gerbert d'Aurillac* (ca. 945-1003), who became Pope in 999. He introduced the Hindu-Arabic numerals on a counting board whose columns represented positive powers of 10, with zero marked by an empty column.

His grasp of this system may have been imperfect, but his efforts heralded the introduction, less than a century later, of new techniques, recently rediscovered from many manuscripts distributed by Arab scribes throughout the regions of the Islamic conquest. A motley group of translators was especially active in the Spanish city of Toledo, which had been retaken by Christian forces from its Moorish rulers in 1085. This military victory provided scholars with access to a multitude of Arabic manuscripts, including translations of Greek scientific and mathematical classics. Many of the earliest Latin versions of these manuscripts were produced via translation from Arabic into Hebrew by scribes from the city's substantial Jewish community, who were fluent in Arabic as well as Latin.

From these local beginnings, translations of Arabic manuscripts obtained from a variety of sources were soon to produce a substantial body of mathematical literature, available in Latin, and widely transmitted to scholars throughout Europe.

2.1. Fibonacci. A key figure in this early period of transmission was *Leonardo of Pisa* (ca.1170 to ca.1250), now more commonly known as *Fibonacci* ('son of Bonaccio'), although this nickname is probably a nineteenth century invention. His *Liber Abaci*, published in 1202, was highly influential. In his youth, Leonardo was taught mathematics in Bugia on the Barbary Coast (now in Eastern Algeria), which was then part of the Western Muslim Empire. He travelled widely throughout the Muslim world, becoming familiar with Euclid's *Elements* and the Greek methods of geometric proof, as well as with the Hindu-Arabic numerals, the decimal place system, and the algebraic approach to solving equations of al-Khwarizmi. Upon his return to Pisa he joined the academic court of the Holy Roman Emperor, Frederick

II, writing several influential texts – of which the practically oriented*Liber Abaci* was by far the most successful and widely read.[3]

Despite its title, the first part of the book focuses on the way in which the Hindu-Arabic numerals provide an *alternative* to mechanical calculation. Since Roman times, practical calculations had usually been performed in Europe with an *abacus* (typically, a wooden frame strung with wires carrying different coloured beads as counters) or similar mechanical device; the final answer was then written down in Roman numerals. Leonardo showed how these mechanical devices could be by-passed by recording on sheets of paper, in Hindu-Arabic notation, the various steps and results of applying simple *algorithms*for combining numerals when adding, subtracting, multiplying or dividing.[4] These algorithms remain essentially unchanged in modern primary school curricula. In other words, *Liber Abaci* became the text for performing calculations *without* the abacus.

Leonardo begins with the numerals:

The nine Indian figures are 9, 8, 7, 6, 5, 4, 3, 2, 1. With these nine figures, and with the sign 0, which the Arabs call **zephir** *(cipher), any number whatsoever is written, as is demonstrated below.*

The bulk of the book is devoted to a wide range of practical problems in mensuration, commerce and currency conversion, as well as developing algebraic techniques to handle a wide range of linear and quadratic equations. In these, and in describing methods for series summation, he gives meticulous geometric justifications of his methods, in the style of Euclid. Notably, in the later chapters he readily uses negative numbers as solutions to certain equations and calculates accurately with these. He justifies rules for adding and multiplying positive and negative numbers, although this is always done in the context of 'practical' calculation, as in the following example, found in Chapter 13 of the *Liber Abaci*.

The problem states that four men find a purse, and that for each, the sum of his original wealth plus the purse, is in a simple proportion in relation to the original wealth of the next two, in a circular pattern: the first plus the purse will have wealth *double the sum of the second and third*, the second plus the purse *triple that of the third and fourth*, the third plus the purse *quadruple that of the fourth and first*, the fourth plus the purse *quintuple that of the first and second*. He shows that this problem can only

---

[3] Astonishingly, this highly influential Latin text was not translated into any modern language for 800 years – an English edition [42] finally appeared in 2002!

[4]The process of producing paper from woodpulp dried into thin flexible sheets was invented in China early in the second century. It was gradually transmitted via the Middle East (especially Baghdad), reaching Western Europe by the thirteenth century, and was soon produced in local paper mills, displacing the earlier uses of papyrus and vellum. The original name for paper in Europe, *bagdadikos*, indicates its tranmission to Europe via the Arabic world.

be solved if *the first man has a debit* (that is, he owes money!) which means that the solution requires negative numbers. The problem is indeterminate in general, but he finds the smallest solution: the four men have original wealth $-1, 4, 1, 4$ respectively, and the purse is 11. These numbers provide a solution, because

$$(-1 + 11) = 10 = 2(4 + 1)$$
$$(4 + 11) = 15 = 3(1 + 4)$$
$$(1 + 11) = 12 = 4(4 + (-1))$$
$$(4 + 11) = 15 = 5(-1 + 4).$$

Leonardo does not write -1 as a number, but expresses it as a 'debt' (what we might call 'negative equity'!).

2.2. The Fibonacci sequence. The final chapter of *Liber Abaci*, and his subsequent *Liber quadratorum* (1225) demonstrate clearly that Leonardo is comfortable with the full range of Islamic algebra, including the solution the general quadratic and certain cubic equations. His handling of now well-known number theory results like the Chinese Remainder theorem (finding numbers that leave pre-assigned remainders when divided by a fixed set of primes) foreshadows *number theory* that would be developed by *Fermat* some 350 years later. Yet he is most widely known for a seemingly much more trivial result, which has fixed his name in modern public consciousness: the *Fibonacci sequence*. This appears innocently in Chapter 12 of the *Liber Abaci*:

A certain man had one pair of rabbits together in a certain enclosed place, and one wishes to know how many are created from the pair in one year when it is in the nature of them in a single month to bear another pair, and in the second month those born bear also.

The (unspoken) assumption here is that the first pair are new-borns at the start of the year in question and that rabbits begin to mate when one month old. This assumption ensures that we treat all the pairs equally. At the end of the first month the first pair is therefore still the only one, but by the end of the second month they have borne their first pair of offspring. The sequence therefore begins with

$$1, 1, 2, 3, 5, 8, 13, 21, 34, 55, 89, 144, ..$$

since, from the third term onward, each term is the sum of the preceding two: by the end of the second month there are two pairs, in the third month the adults produce another pair (so now there are 3 pairs), in the fourth month both they and their firstborn produce pairs (making the total 5 for the fifth month) and so on. By the end of the twelfth month (assuming all the rabbits survive till then) we have 144 pairs.

While this is an old and rather trivial problem—the sequence appears in various Indian mathematical writings as early as the sixth century—it gains mathematical interest when one considers the successive ratios:

$$\frac{1}{1}, \frac{2}{1}, \frac{3}{2}, \frac{5}{3}, \frac{8}{5}, \frac{13}{8}, \frac{21}{13}, \frac{34}{21}, \dots$$

These ratios can be shown to come ever closer to the *golden ratio* $\frac{1}{2}(1 + \sqrt{5})$. The term "golden ratio" seems to have been coined as late as the nineteenth century. However, the number itself generated much excitement in artistic circles, especially during the Romantic era.[5]

To see how this limiting ratio is found, we denote the nth Fibonacci number by f_n. Hence the Fibonacci sequence starts with $f_1 = f_2 = 1$, and $f_{n+1} = f_n + f_{n-1}$ for $n = 2, 3, \dots$. The ratios $(r_n)_{n \geq 1}$ of successive terms satisfy the identity

$$r_n = \frac{f_{n+1}}{f_n} = \frac{f_n + f_{n-1}}{f_n} = 1 + \frac{f_{n-1}}{f_n} = 1 + \frac{1}{r_{n-1}}.$$

If we accept (for a proof see *MM*) that the $(r_n)_n$ do indeed *converge* ('get ever closer to') to some number x as n grows, then the limiting value x of the two sequences $(r_n)_{n \geq 1}$ and $(r_{n-1})_{n > 1}$ must clearly be the same. Therefore the relationship $r_n = 1 + \frac{1}{r_{n-1}}$ will, for large n, approximate the equation $x = 1 + \frac{1}{x}$, which we can write as $x^2 - x - 1 = 0$. The usual formula for the solution of a quadratic equation provides the positive value

$$x = \frac{1 + \sqrt{5}}{2} = 1.6180339987\dots$$

Since we know from the Theodorus lesson that $\sqrt{5}$ is not a rational number, it follows that neither is x, the golden ratio. This again raises the question how one should *define* these quantities consistently.

In geometric terms, the 'golden ratio' (also called the 'golden section') has a history that predates Leonardo by at least 1500 years. Greek geometers (possibly even in Pythagorean times) were concerned to determine the point C on a straight line segment AB such that 'the whole is to the greater part as the greater is to the smaller'. Euclid called this the division of AB in *mean and extreme ratio*. By this he meant that the lengths AC, BC should satisfy the proportion

$$AB : AC = AC : BC,$$

(which makes AC the *mean proportional* between AB and BC, as discussed in Pythagorean music theory) and finding C requires the square root of their product, as

$$AC^2 = AB.BC.$$

---

[5]More recently, much has been published about the apparent ubiquity of the Fibonacci sequence and the golden ratio in nature, be it in the shapes of snail shells, configurations of flower petals, seed heads, pine cones, etc. A guide to the underlying mathematics of these phenomena can be found in the classic text *Introduction to Geometry* by HMS Coxeter. A less technically challenging reference is *Fascinating Fibonaccis* by Trudi Hammel Garland.

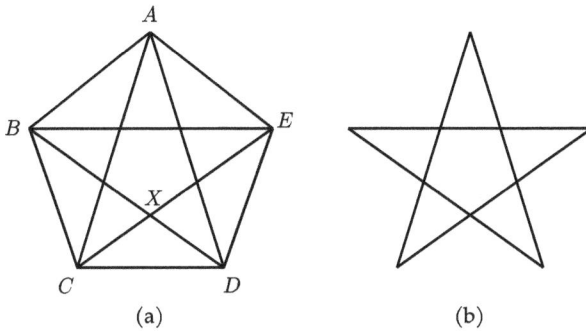

Figure 13. Pentagon and pentagram

Geometrically, AC is the side of a *square* whose area is the same as that of the *rectangle* with sides AB, BC.

To find C we put the 'greater part' $AC = a$, the 'lesser part' $BC = b$, making 'the whole' $AB = a + b$. The mean and extreme ratio is then

$$\frac{a+b}{a} = \frac{a}{b},$$

so it satisfies $\frac{a}{b} = 1 + \frac{b}{a}$. Setting $x = \frac{a}{b}$, this produces the identity $x = 1 + \frac{1}{x}$ satified by the limit of the Fibonacci ratios, so that $\frac{a}{b} = \frac{1+\sqrt{5}}{2}$. Remember, however, that $\frac{a}{b}$ is a ratio of *lengths* and that a, b cannot be taken as multiples of a chosen unit, since this identity has no solution in whole numbers.

The golden section is easy to construct geometrically. It arises as the ratio between the diagonal and side of a *regular pentagon*, i.e. a pentagon whose sides and internal angles are all equal. This fact suggests that its origins may well lie in Pythagorean times. The regular pentagon and the pentagram – the five-pointed star comprising the diagonals, which some claim may have been used by the Pythagoreans as a sign of recognition – see [25]) are pictured in Figure 13.

In Figure 13(a) the regular pentagon $ABCDE$ has diagonals AC, AD, BD, BE and CE, and if X is the point of intersection of BD and CE, the triangles BDC and CDX are similar, so that the corresponding sides are in proportion: for example, the ratios of lengths $\frac{BD}{CD}$ and $\frac{CD}{XD}$ are equal. Hence $CD^2 = BD.XD$. But clearly $CD = BC = BX$, so $BX^2 = BD.XD$ so that X divides BD in mean and extreme ratio. The pentagram is shown in Figure 13(b).

2.3. Maestri d'abbaco. Leonardo was clearly a highly accomplished mathematician, yet, apart from occasional glimpses of originality, his work is

chiefly a well-rounded compilation of earlier discoveries by the Greeks, Hindus and Arabs. In bringing these different strands together, however, he and
the other translators of the twelfth and thirteenth century stimulated the revival of mathematics in Europe, especially in Italy, France and (a little later)
in England and in the many Germanic statelets of Central Europe. While the
tensions created by problems whose solutions included negative numbers
and roots of equations that could not be expressed as ratios of whole numbers became increasingly apparent, it would take several generations before
mathematicians were fully persuaded of the need to accept such solutions
as *numbers*.

Through the success of Leonardo's *Liber Abaci*, the discipline (*abbaco* in
Italian) of replacing the abacus and Roman numerals with the algorithms
for calculating with Hindu-Arabic numerals, soon spread widely throughout much of Western Europe, especially through the growing influence of
mercantile cities like Pisa, Genoa, Venice and Amalfi. In place of barter and
direct trade, the international trading companies of these cities employed
more sophisticated practices such as bills of exchange, promissary notes, letters of credit and loans, all of which required double-entry bookkeeping and
arithmetical skills that were *not* part of the quadrivium taught in the early
universities of Bologna, Paris or Oxford. The problem-solving techniques
displayed in the *Liber Abaci* led to a new cadre of mathematics teachers, who
became known as *maestri d'abbaco*, meeting the needs of the merchant class.
Pen and paper calculation based on the decimal system began to replace
the traditional counting boards, and a large collection of problem-solving
manuals were produced to instruct budding merchants.

The innovations these *maestri* introduced were largely notational: the
Arabs and Leonardo had described their calculations using words rather
than symbols, but the pressure of teaching meant that, for the *maestri*, abbreviations (for square, cube, etc) soon became a common practice. The
symbols for unknowns we find so familiar were still some way off, however. Typically, an unknown (which we might denote by x) was known as
'the thing' (*cosa*), and what we would write as $x + \sqrt{y}$, a fourteenth century instructor would typically describe as *a thing plus a root of some quantity*. An 'equation' might be constructed verbally, involving the unknown
and its square and some numbers, and manipulated, according the rules already described by al-Khwarizmi, into one of his six standard forms. Gradually, problems being considered would include some aspects that seemed
divorced from practical (commercial) use. The increase in complication in
the expressions being explored led to the use of abbreviations, such as c for
the 'cosa' or unknown, ce for its square, 'censo', and cu for its cube, 'cubo'.
Higher powers, e.g. $ce\ cu$ ('censo di cubo') also began to appear. At this
stage these abbreviations remained just that: no apparent meaning as *numbers* was attached to them.

The introduction of further symbolic notation was prominent in the work of the German *Coss* tradition, which had adapted the practices of the Italian *maestri* by the early sixteenth century. The influential treatise *Die Coss*, published in 1525 by *Christoff Rudolph* (1499-1545) contains a full list (in Germanic script) of symbols for powers up to the ninth and makes full use of signs such as $+$ and $-$ as well as a square root sign, to shorten the complicated expressions he discusses. (See [6] for a more detailed discussion.)

In practice, the consistent use of symbols helped to speed the acceptance of quantities that lacked geometric representation. However, Rudolph, who clearly recognised the link between what we would call the arithmetic progression $0, 1, 2, 3, 4, ...$ and its geometric counterpart $1, 2^1, 2^2, 2^3, 2^4...$, did not take the seemingly obvious next step of using the symbol 2^k for the kth member of the latter progression.

This advance had already been made in France by *Nicolas Chuquet*, whose precise dates are somewhat uncertain. He was possibly the first to use zero and negative numbers as exponents. His unpublished 1484 manuscript, *Triparty en la Science de Nombres* (rediscovered only in the late nineteenth century) was partly reproduced, without acknowledgement, in a 1520 textbook by *Etinenne de la Roche*. Chuquet invented an essentially modern notation to describe arbitrary integral powers and also (as we saw in the **Prologue**) introduced names for ever higher powers of 10. He used positive exponents for positive powers of the unknown, such as $.8.^3$ (which we would read as $8x^3$) and denoted negative exponents by adding m after the exponent, as in $.7.^1m$, which is our $7x^{-1}$. He recognised that multiplication of these two terms involves multiplication of the coefficients and *addition* of the exponents, resulting in $.56.^2$ (in our terms, $56x^2$). For addition and subtraction he employed the abbreviations also used by *Luca Pacioli* (1445-1517): $\overline{p}, \overline{m}$ for 'plus' and 'minus'. Our 'equals' sign $=$ first appeared in the 1557 *Whetstone of Witte* by *Robert Recorde* (1512-1558), who also introduced $+, -$ to Britain.

Throughout the sixteenth century, the boundaries between the arithmetical techniques inspired by the *maestri* and the study of the Greek classics at universities gradually became blurred. Newer editions of Euclid's *Elements* began to utilise arithmetical descriptions of Euclidean results, especially in Book X, where Euclid classifies various types of incommensurables (see **Chapter 1**, Section 4). His difficult geometric constructions become more transparent if the outcomes are presented as surds. Thus calculation using rationals and irrationals together no longer appeared as foreign. Calculations with surds such as $(2 - \sqrt{3}) \times (2 - \sqrt{3})$ also drew attention to the need for consistent rules when multiplying signs. Formal rules for sign multiplication featured prominently in Pacioli's widely-read 1494 publication, *Summa de arithmetica geometria proportioni et proportionalita*. This contained a systematic account of the techniques pioneered by the *maestri* and *cossists*, pointing to their practical as well as theoretical utility.

Despite these advances, resistance to the acceptance of the reality of negative solutions to problems as *numbers* proved more difficult to overcome. I pick up this story in the next chapter, but will turn first to further progress in the solution of equations.

3. Solving equations: cubics and beyond

While progress with algebraic symbolism was steady rather than spectacular, in the first half of the sixteenth century algebraic techniques for solving polynomial equations beyond the quadratic took a major step forward. Publicly, the catalyst was the publication of the *Ars Magna* (1545) by *Girolamo Cardano* (1501-1576). The story behind this highly influential book is quite convoluted. The breakthrough was probably made in the early 1520s, but remained buried in secrecy for two decades. An unfortunate consequence was that the mathematician primarily responsible for advancing the subject at this time, *Scipione del Ferro*, died in 1526 and was largely forgotten by his peers and in many subsequent historical accounts. This has also meant that no direct evidence remains of the process by which he arrived at his results, although the final version of the techniques is well-documented.

For some time after the work of Leonardo of Pisa, the analysis of cubic equations had remained firmly within the Arabic tradition of addressing the problem geometrically, rather than algebraically. This tradition boasted a major work by *Omar Khayyam*, the polymath perhaps best known for a selection of about 1000 poems given the title *The Rubaiyat* by its Victorian translator, Edward FitzGerald.

Omar Khayyam's best-known mathematical treatise, *On the Proofs of the Problems of Algebra and Muqabala*, classifies fourteen different types of 'cubic equation' (he actually lists 25 cases, but 11 of these reduce to become linear or quadratic) with *positive* solutions. In each case he shows how to construct the solution of the cubic equation geometrically. His solutions invariably require the construction of a *conic section*. Specifically, this is either a circle, an ellipse, a parabola or a hyperbola, formed by the intersection of a plane with a circular cone (see Figure 14 in **Chapter 3**). The last three cannot be constructed by straightedge and compass alone—we will discuss their Ancient Greek origin in the next chapter.

Omar Khayyam insists that powers higher than cubes, such as the *square-square* of Diophantus, do not exist 'in reality' although he states that they constitute 'theoretical facts'. He argues that algebraic methods can be used to solve cubic equations, but that proofs dealing with the third power require solid geometry.

Once again it is much more convenient to outline his methods in algebraic terms. Today we write the general cubic equation in the form

$$x^3 + ax^2 + bx + c = 0,$$

since we can always divide by the (non-zero) coefficient of x^3 if we need to.

However, Khayyam worked geometrically and insisted on writing the coefficient of the linear term in x as a square and the constant term in the equation as a cube: this would enable him to maintain the 'three-dimensional nature' of each of the terms on the left. The equation is therefore written as

$$x^3 + ax^2 + b^2x + c^3 = 0.$$

We express his method in modern notation: the substitution $x^2 = 2py$ (which defines a *parabola* centred at the origin) turns the equation into

$$2pxy + 2pay + b^2x + c^3 = 0,$$

which Khayyam recognises as defining a *hyperbola*. He has to deal separately with different cases for a, b, c, any of which can be positive, negative or zero, and he restricts himself to cases where the solution (the point of intersection of the parabola and hyperbola in question) has positive coordinates. While he works entirely in a geometric setting, he treats his solutions, whether rational or irrational, as providing *numbers*, rather than as geometric magnitudes.

3.1. Cardano's formula. This, more or less, remained the state of affairs for cubic equations as inherited by *Scipione del Ferro* around 1500. In Italy, as familiarity with decimal notation and algebraic manipulation grew, the problem of finding the solutions of various classes of cubics by means of an *algebraic* formula, rather than through a geometric construction, had begun to receive much attention. At this time also, the solution of particular equations – still expressed rhetorically – lent itself to a kind of public mathematical jousting which gradually became a popular sport in some cities. Scholars would challenge each other to solve problems in public, with bystanders engaging in bets on the outcome. Skilled combatants saw these challenges, and the associated betting, as a way of making a living from their craft.

In one such contest in 1535, *Niccolo Fontana* (1499-1557), an able scholar nicknamed *Tartaglia* (The Stammerer) roundly defeated the less talented *Antonio Fior*. It seems that the latter had learnt how to solve one class of cubic equations when studying as a pupil of del Ferro. As became apparent somewhat later, del Ferro had probably managed to find solution methods for all three classes into which cubics with positive solutions had been divided at the time. Tartaglia, possibly aware of del Ferro's success, had managed, independently, to solve one class of cubic equations. In a short burst of intense creative activity just before the contest, he was also able to master the solution method for the class of equations Fior had learnt from del Ferro. He now set problems that Fior could not solve, while solving all thirty equations that Fior had set for him. The contest caused a minor sensation.

Hearing of this, Cardano approached Tartaglia, and, after much effort, persuaded the reluctant 'Stammerer' to divulge his techniques. Although Tartaglia, wishing to protect his livelihood, had sworn him to absolute secrecy, Cardano was already in the process of constructing his *Ars Magna* as a definitive algebra text and decided to break his promise. He tried to make amends by clearly citing del Ferro and Tartaglia as the authors of the solution methods for cubics featured in the book.

However, when the *Ars Magna* appeared, a furious Tartaglia accused Cardano of plagiarism and treachery. It was too late to repair the damage. A bitter dispute ensued, in which Cardano's principal defence was that the late del Ferro, rather than Tartaglia, had been the first to solve all three types of cubic equation. Cardano and his gifted student *Lodovico Ferrari* (1522-1565) claimed that they had confirmed this when they consulted del Ferro's papers, kept by yet another of del Ferro's students in Bologna, in 1543.[6]

Although Tartaglia published his book *Quesiti et Inventione Diverse* the following year, setting out his methods in verse, in coded form, Cardano's clear exposition and comprehensive range, presented in his *Ars Magna*, had already brought him fame that overshadowed Tartaglia's.[7] In the ensuing controversy, Cardano was ably supported by Ferrari, who had by then discovered a general solution method for quartic equations (see below), which Cardano also included in his *Ars Magna*. In a much-delayed challenge contest with Tartaglia in 1548, Ferrari energed a clear victor. An embittered Tartaglia died in obscurity and poverty nine years later.

One can appreciate the scale of the advances made by del Ferro and Tartaglia by reading Cardano's text. As usual, it is much easier for us to understand the method using modern algebraic notation. The wording of Cardano's verbal solution is reminiscent of solutions described on Babylonian clay tablets or by al-Khwarizmi. Here is an extract for comparison before we discuss his solution methods in modern terminology.

Cube the third part of the number of 'things' , to which you add the square of half the number of the equation, and take the root of the whole, that is square root, which you will use, in the one case adding the half of the number which you just multiplied by itself, in the other case subtracting the same half, and you will have a 'binomial'

---

[6]Extensive extracts from the correspondence between all these combatants are presented in [12], including the extract from Cardano's *Ars Magna* reproduced below.

[7]Tartaglia's book is available at http://www.it.wikisource.org/wiki/Quesiti_et_inventione_diverse. See also [6]

and 'apotome' respectively; then, subtract the cube root of the apotome from the cube root of the binomial, and the remainder of this is the value of the 'thing'.

Note the use of the Euclidean terms 'binomial' and 'apotome' (see **Chapter 1**, Section 3) to describe the sum and difference of the two terms under the cube root. Cardano provides an example, using the cubic $x^3 + 6x = 20$. He states this equation verbally as: *'the cube and 6 'things' equals 20'*, and leaves his solution in the form

$$\sqrt[3]{(\sqrt{108} + 10)} - \sqrt[3]{(\sqrt{108} - 10)}$$

which he does not attempt to simplify further, although $x = 2$ clearly also solves the equation.

Even in modern notation, the algebraic manipulations we need to arrive at Cardano's solution are somewhat more technical than those used so far – readers in a hurry may skip the two shaded sections below without much loss of continuity.

The general cubic equation

$$x^3 + ax^2 + bx + c = 0$$

can be reduced to an equation of the form $y^3 + py = q$ for appropriate constants p, q, so that the quadratic term is elimininated (following a pattern encountered already in the work of Diophantus). Simply set $y = x + \frac{a}{3}$, so that $x = y - \frac{a}{3}$. Expressed in terms of y, the equation takes this form, with $p = b - \frac{a^2}{3}$ and $q = -(c + \frac{2a^3}{27} - \frac{ab}{3})$.

To solve $y^3 + py = q$, first notice that for *any* A, B we have

$$(A - B)^3 = A^3 - 3A^2B + 3AB^2 - B^3,$$

so

$$(A - B)^3 + 3AB(A - B) = A^3 - B^3.$$

Thus, if we can find A, B to satisfy $3AB = p$ and $A^3 - B^3 = q$, then $y = A - B$ satisfies $y^3 + py = q$, and $x = y - \frac{a}{3}$ solves the original equation.

To find such A, B for the given values of p, q, note that $B = \frac{p}{3A}$ will lead to $A^3 - \frac{p^3}{(3A)^3} = q$, which reduces to a *quadratic* equation in A^3, namely

$$(A^3)^2 - qA^3 - (\frac{p}{3})^3 = 0.$$

The formula for the general quadratic now provides the solutions

$$A^3 = \frac{q}{2} \pm \sqrt{(\frac{q}{2})^2 + (\frac{p}{3})^3}$$

(although Cardano only takes the positive square root) and therefore

$$B^3 = -\frac{q}{2} \pm \sqrt{(\frac{q}{2})^2 + (\frac{p}{3})^3}.$$

Finally, Cardano writes

$$y = A - B = \sqrt[3]{\frac{q}{2} + \sqrt{(\frac{q}{2})^2 + (\frac{p}{3})^3}} - \sqrt[3]{-\frac{q}{2} + \sqrt{(\frac{q}{2})^2 + (\frac{p}{3})^3}},$$

which now allows him to solve the original cubic by substituting the above values for p, q, and using $x = y - \frac{a}{3}$.

Somewhat ironically, this has become known as *Cardano's formula* (although some authors remember to credit del Ferro and/or Tartaglia). From Cardano's perspective, moreover, the formula only deals with one of the 'cases' he considered, namely $y^3 + py = q$, since he wishes to avoid negative coefficients at any stage.

This means (for example) that, in order to deal with the case $y^3 = py + q$ for positive p, q, he cannot simply write $y^3 - py = q$ and use the above argument with $-p$ instead to arrive at the solution! Instead he makes an elaborate substitution, based on the identity $(A+B)^3 = A^3 + B^3 + 3AB(A+B)$, which he justifies geometrically. We will soon discover why in such cases his formula would cause him major conceptual difficulties of a different kind.

As noted above, a general solution method for the general *quartic* (fourth-degree) equation of the form

$$x^4 + ax^3 + bx^2 + cx + d = 0,$$

found by *Ferrari* in 1540, was also included in Cardano's *Ars Magna*.

In brief, this solution method proceeds via the substitution $y = x - \frac{1}{4}$ to turn the general quartic equation into the form $y^4 + py^2 = -qy - r$, for appropriate new coefficients p, q, r which can again be expressed in terms of a, b, c, d. To attack the reduced equation, Ferrari notes that in the perfect square $(y^2 + \frac{p}{2})^2 = y^4 + py^2 + \frac{p^2}{4}$, the first two terms are what we have on the left in the above, so that $(y^2 + \frac{p}{2})^2 = \frac{p^2}{4} - qy - r$. He now adds a further unknown z to $(y^2 + \frac{p}{2})$ on the left, and again computes the square, substituting $(y^2 + \frac{p}{2})^2$ as above:

$$((y^2 + \frac{p}{2}) + z)^2 = 2zy^2 - qy + (z^2 + pz + \frac{p^2}{4} - r)$$

On the left we have expressed this quantity as a quadratic in y, and as it is a perfect square, so we need to find the value of z such that the discriminant of this quadratic is 0.

Recall that the *discriminant* of the general quadratic $ax^2 + bx + c = 0$ is $b^2 - 4ac$. In the present case we have $a = 2z, b = -q, c = (z^2 + pz + \frac{p^2}{4} - r)$, so that

$$q^2 = 4(2z)(z^2 + pz + \frac{p^2}{4} - r).$$

This cubic equation in z can be solved by use of Cardano's formula. Let $z = z_0$ be such a solution. For this value of z, the above quadratic in y has a double root $y_0 = \frac{q}{4z_0}$ (since the discriminant is zero, the root is $\frac{-b}{2a}$ in general). But now our equation in y, z above has the form $((y^2 + \frac{p}{2}) + z)^2 = 2z_0(y - y_0)^2$. So, for these values, the reduced quartic splits into a pair of quadratics whose solutions form the roots of the equation[a].

[a]This outline is based on: Ferrari method. Encyclopedia of Mathematics. URL: http://www.encyclopediaofmath.org/index.php?title=Ferrari_method&oldid=35675

And there, despite continuing efforts, matters would rest for more than 250 years. No-one was able to produce a formula similar to the ones described above in order to find the roots of polynomials of degree 5 (quintics) or higher. Nonetheless there was significant, if gradual, progress in the general understanding of the structure and theory of polynomial equations throughout the 17th and 18th centuries.

In 1799 the youthful German mathematician *Carl Friedrich Gauss* (1777-1855) asserted, without giving a proof, that the quintic has *no general solution formula* by means of radicals (i.e. using roots as was done above). Between 1799 and 1813 the Italian medical doctor, philosopher and mathematician *Paolo Ruffini* (1765-1822) published six versions of what he maintained was a proof that polynomial equations in powers higher than 4 have no such solution formulae, although his proofs were opaque and all contained significant flaws.

The brilliant young Norwegian mathematician *Niels Abel* (1802-1829), read Ruffini's work as a student and recognised that it was incomplete. However, by 1824 he had found a rigorous proof of Gauss' claim: there is no single *general formula* which will yield the solution of every quintic. Abel travelled to Paris, then very much the leading centre for algebra, hoping to develop his methods further by collaborating with the leading French mathematicians of the time. However, he found them unresponsive and soon labelled them as *'monstrous egotists'*, unwilling to collaborate with each other and especially with foreigners!

The brilliance of Abel's mathematical insights was not recognised during his short life. He suffered poverty and ill health, unable to find a university position. Just as others began to appreciate the outstanding merit of his work, he died of tuberculosis, aged 26, having by then turned away from polynomial equations. His work in other fields, especially in what are now known as elliptic and 'abelian' functions, was far-sighted and led to major advances in both number theory and algebra.

Three years later, modern algebra was revolutionised by the tragic, irascible French prodigy *Èvariste Galois* (1811-1832), who, as a left-wing firebrand, was mortally wounded at the tender age of 21 in a duel. Galois' methods explained, as a by-product of a much broader algebraic theory now named after him, why the general quintic and higher-order polynomial equations can have no such solution formula. The technical details are well beyond the scope of this book: a description of these events is given in [35].

3.2. Imaginary roots. The renown of the work of Cardano and his compatriots led to much greater interest in algebraic techniques from the late sixteenth century onwards. Cardano continued to describe negative numbers as 'fictitious' (*numeri ficti*) and ignored them when they occurred in his formula. He also encountered difficulties when solving equations of the form $y^3 = py + q$, since in that case the quantity under the square root sign in Cardano's formula is $(\frac{q}{2})^2 + (\frac{-p}{3})^3$, which can become negative for particular values of p and q.

Consider, for example, the cubic equation $y^3 = 15y + 4$, which clearly has the solution $y = 4$. Cardano's formula for this equation yields

$$y = \sqrt[3]{2 + \sqrt{-121}} + \sqrt[3]{2 - \sqrt{-121}}.$$

Cardano could not deal with the term $\sqrt{-121}$, since square roots of negative numbers seemed to make even less sense than negative numbers themselves! He could not understand why his formula would not yield the obvious root $y = 4$. He had noticed that a similar problem arises with quadratic equations: in examples of the form $x^2 + b = ax$ he realised that the usual solution formula would involve the square root of a negative number if $a^2 < 4b$. An example of this kind, included in his later work *Ars magnae sive de regulis algebraicis liber unus*, has $a = 10, b = 40$. He writes this equation as $x(10 - x) = 40$. He proposes $5 + \sqrt{-15}$ and $5 - \sqrt{-15}$ as solutions, but still regards these as 'impossible', while conceeding that they are 'operationally' correct. He regards $\sqrt{-15}$ as meaningless, calling it a 'quantita sophistica'.

Two decades later, *Rafael Bombelli* (1526-ca.1572), aware that the cubic equation $y^3 = 15y + 4$, which had so troubled Cardano, has the solution 4, had what he called a 'wild thought'. Perhaps, he argued, one could make sense of 'numbers' of the form $a + b\sqrt{-1}$ (for $2 + \sqrt{-121}$ take $a = 2$ and $b = 11$) by setting out *multiplication tables* for $\sqrt{-1}$, similar to those already in use for positive and negative numbers – although even these had not yet been properly justified! By analogy with what he knew from working with $+$ and $-$ for integers, including, notably $(-1)(1) = -1$ and $(-1)(-1) = 1$, he argued that the square root of a negative number 'has different arithmetical operations from the others'.

These analogies led him to propose new multiplication rules that we would today write as:

$$(\sqrt{-1})(\sqrt{-1}) = -1 = (-\sqrt{-1})(-\sqrt{-1})$$
$$(-\sqrt{-1})(\sqrt{-1}) = 1 = (\sqrt{-1})(-\sqrt{-1}).$$

Bombelli articulated these relationships by means of verbal expressions, using *piu di meno* for $\sqrt{-1}$ and *meno di meno* for $-\sqrt{-1}$. For example, he expressed his key new rule (which we write as $(\sqrt{-1})(\sqrt{-1}) = -1$) as *piu di meno via piu di meno fa meno*.

Having experimented with such 'multiplication tables' for the square roots of *numeri ficti*, he proceeded to apply these in solving the troublesome equation $y^3 = 15y + 4$. Taking the cube of each term in the Cardano formula $y = \sqrt[3]{2 + \sqrt{-121}} + \sqrt[3]{2 - \sqrt{-121}}$, he obtained the (verbal) equivalent of

$$2 + \sqrt{-121} = (a + b\sqrt{-1})^3$$
$$2 - \sqrt{-121} = (a - b\sqrt{-1})^3$$

for some unknowns a, b. Multiplying out the perfect cubes on the right and using his new multiplication tables for $\sqrt{-1}$, he found (see *MM* for the calculation) that the choices $a = 2, b = 1$ provide a solution. Using this in the Cardano formula, Bombelli obtained the positive integer solution $y = (2 + \sqrt{-1}) + (2 - \sqrt{-1}) = 4$ for the equation! Thus, Bombelli's multiplication tables, used in Cardano's formula, produced the positive solution that was apparent by direct inspection, even though the terms of the formula appeared to include quantities that Cardano had considered to be 'pure fictions'.

Although he did not claim that the square root of a negative number should be accepted as a *number*, Bombelli had shown that using his rules for the multiplication of such quantities gave the correct positive root of this troublesome cubic equation. Bombelli was not able to make logical sense of his discovery, but he had in fact derived the correct multiplication rules for working with $\sqrt{-1}$, which we today call the *imaginary unit i* in the complex plane, to be discussed in **Chapter 4.** His calculation, while remaining speculative, did not banish the suspicion that such quantities had aroused, but perhaps it served to diminish it somewhat.

CHAPTER 3

Construction and Calculation

But the best demonstration by far is experience, if it go not beyond the actual exper-
iment.

Sir Francis Bacon, *Novum Organum*, 1620

Summary

This chapter takes a step back, to review approaches to geometric construction in Classical Greece. This sets the scene for three influential problems (doubling the cube, trisecting the angle and squaring the circle), whose
solutions had proved elusive under the constraints of Euclidean geometry. We consider ingenious solutions from Ancient Greece that led to novel
curves and reformulate two of these problems in terms of cubic equations
(in one case by means of basic trigonometry), before moving on to Europe
around 1500, to pick up the further development of the decimal system
as well as the development of aids to calculation. These were needed for
lengthy calculations in astronomy, where trigonometric tables had long been
used, and increasingly also in navigation. The chapter closes with a brief account of *John Napier*'s invention of logarithms.

1. Constructions in Greek geometry

While Euclid's *Elements*, and the work of many of his successors, restricted geometric constructions to those that can be completed by means of
a *straightedge* (an unmarked ruler) and *compass* alone, it was soon discovered
that the construction of certain lengths and angles with these tools alone
posed seemingly insuperable difficulties. Consequently, Greek geometers
invented various ingenious devices for the purpose of drawing curves that
could provide the solutions they sought, however much such techniques
might offend purists by stepping outside the strict rules for geometric construction set out definitively in Euclid's *Elements*.

One of these new solution methods led directly to the invention of the
conic sections (ellipse, parabola and hyperbola), so named because they are
found when cutting a (double) circular cone with planes at various angles to

its axis. These curves were to become indispensible tools in the development of modern mathematics and its applications to physics and cosmology.

We may think of an upturned ice cream cone in order to imagine a circular cone. Its *axis* is the line connecting its vertex to the centre of the circle at the base of the cone, and a *generator* is any surface line connecting a point on the circumference of the base circle with the vertex – since rotating this line through a full revolution about the axis will sweep out the cone. As displayed in Figure 14(a), cutting the cone by a plane parallel to the base produces a *circle*, while (as in Figure 14(b)), cutting it by a plane having an angle to the base less than that of a generator produces an *ellipse*. Both are closed curves (i.e. their ends meet). But if the plane is parallel to a generator, as in Figure 14(c), we obtain a *parabola* (an open curve). If the angle is even steeper, then, as in Figure 14(d), we need a double cone with a common vertex (e.g an ice-cream cone whose vertex rests on a mirror, together with its mirror image), and the *hyperbola* we obtain has two (open) branches, one in each cone.

The culmination of Greek geometers' analysis of these curves has been preserved for us in translations of the impressive *Treatise on Conic Sections* by *Apollonius of Perga* (ca 240-190 BCE), known to his peers as 'The Great Geometer'.

After the Renaissance, European mathematicians, scientists and architects found both theoretical and practical uses for the different conic sections in a wide variety of areas. *Johannes Kepler* (1571-1630), after a close study of astronomical measurements made by *Tycho Brahe* (1546-1601), postulated correctly that planetary orbits are not circular, as had been supposed, but elliptical, and the Astronomer Royal, *Edmund Halley* (1656-1742), applied this in 1682 to other bodies in orbit around the sun, predicting correctly that the famous comet named after him would return in 1758. *Galileo Galilei* (1564-1642) argued that the path of a projectile fired from a cannon would follow a parabola.

Today, parabolic mirrors creating parallel beams of light are used in car headlamps. A striking example of a hyperbolic shape is given by the cooling tower of any power station generating steam. On a similarly practical level, some 'whispering galleries' exploit a property of the two focal points of an ellipse: a person near one of the focal points can clearly hear sounds generated near the other focal point.[1]

---

[1]An ellipse can also be defined as the shape traced in a plane by holding the sum of its distances from two given points (its *focal points*) fixed. (Take a piece of string, longer than the distance between these two points, on a sheet of paper, fix the endpoints of the string to these points, hold the string taut inbetween with a pencil point and draw an arc all the way round - this is your ellipse.)

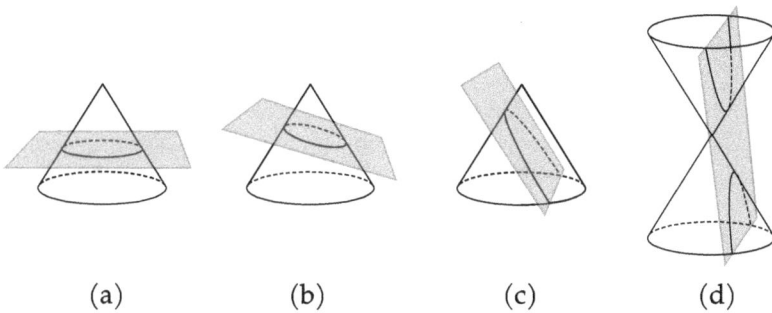

(a)   (b)   (c)   (d)

Figure 14. Conic Sections

Constructions using a marked ruler – known as *neusis* constructions, after a Greek term meaning 'inclined towards'– also proved effective in attacking problems that were difficult or impossible to solve by other means. Here certain lengths (e.g. one or more multiples of a chosen unit) can be marked off on the ruler, which is then used to fit a line segment of specified length between two given lines or circular arcs, while also going through some given point (i.e. the specified segment should 'point in a given direction').

Some preference for straightedge-and-compass construction over these other construction techniques is said to have been expressed by early geometers in the fifth century BCE, and during the fourth century BCE this opinion gained ground due to the influence of Plato's philosophy. Plato taught that physical appearance, as observed by our senses, is merely a 'shadow' mimicking the real 'Form' of the object in question. He proposed this view as an answer to the *problem of change* posed by pre-Socratic philosophers of the sixth and fifth century BCE. *Parmenides* insisted that what is, has always been so, and that, logically, nothing can ever change, since it is impossible for something to come out of nothing. On the other hand, *Heraclitus* insisted that the basic essence of the universe is ever-present change ('no man ever steps in the same river twice'), and thus (as Plato later quoted Heraclitus) 'all entities flow and nothing remains still'.

To resolve these conflicting views Plato argued that his 'Theory of Forms' could explain the 'true nature' of things. He maintained that our senses do not allow us a clear view of reality (an assertion perhaps influencing St Paul's famous phrase, four centuries later, that we see 'through a glass darkly') We should therefore reject the essential reality of the physical world. Plato exhorted his followers to study the Forms through mental contemplation as the perfect, unchanging essences of objects, not limited by space or time. He distinguishes between knowledge, which is certain and unchanging, and opinion, which is changable and derives from our illusory sense experience.

The Forms are the source of our innate knowledge, and learning is the process of bringing it to the surface. He frequently argues that mathematical reasoning can facilitate this process and provide a bridge between the physical world and his 'world of ideas' where the Forms exist. Points, lines and circles are therefore to be seen as ideal objects, whose physical representation is a crude approximation of their real nature, which alone should be the object of their study. He criticised geometers whose constructions departed from the aim of mirroring the ideal objects by using 'physical' devices.

Euclid's *Elements* would probably have escaped Plato's strictures, if Plato had survived to read this work. The *Elements* featured rigorous constructions and several conceptual innovations now attributed to Plato's one-time student Eudoxus. Euclid's *axiomatic approach* placed emphasis on keeping the initial collection of definitions and (unproven) postulates as small and as 'self-evident' as possible. In keeping with this approach, physical instruments for drawing figures should be as simple as possible. The straightedge and compass are the only tools Euclid allows in his many constructions.

It seems therefore that Greek geometers recognised a hierarchy of acceptable techniques for geometric constructions: a given problem should first be attacked solely by Platonic methods; if that failed, a construction that also used conic sections might be deemed acceptable. If neither of these approaches produced a satisfactory solution, then attacks on the problem using neusis constructions or other (often quite specific and complicated) 'mechanical' tools might be deemed acceptable.

However, every individual's influence wanes over time. Nearly two centuries after Plato, Apollonius wrote his definitive treatise on conics – which, 1800 years later, proved to be of critical importance in the study of motion and calculations of planetary orbits. A few decades earlier, Archimedes had freely made use of neusis constructions in his geometry and its many practical applications, as well as devising an entirely novel 'method of discovery' (see **Chapter 5**), using physical principles and postulating the infinite divisibility of figures, in order to determine the relationships between the areas and volumes of different curvilinear figures, including conic sections and spirals.

2. 'Famous problems' of antiquity

Three construction problems from antiquity stand out as having inspired major advances in mathematics, as well as helping to create conditions forcing mathematicians to widen their concept of number in order to arrive at satisfactory solutions. These became known as the *three famous problems* of antiquity: duplication of the cube, trisection of the general angle, and squaring the circle. We begin with a brief overview of the first two, deferring discussion of the third, whose solution proved to be rather more involved.

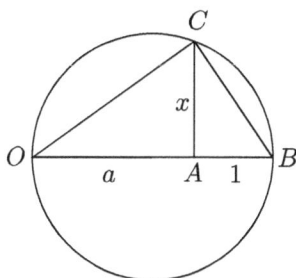

Figure 15. Construction of a mean proportional

2.1. Doubling the cube. Once the Pythagoreans had solved the problem of 'doubling the square' by using its diagonal as the side of a new square with double its area, an obvious question was how to double the cube. The doubling of the square of side a involved finding the mean proportional between a and $2a$, since the relation

$$\frac{a}{x} = \frac{x}{2a}$$

becomes $x^2 = 2a^2$, which has solution $x = \sqrt{2}a$. This therefore involved the construction of the irrational length $\sqrt{2}$, easily done by drawing the diagonal of the unit square.

Moreover, early Greek geometers were well aware that *any* square root has an easy straightedge-and-compass construction, illustrated in Figure 15, which became a staple of schoolbooks through two millennia.

So it was natural to ask how such methods might be extended to generate cube roots. It was observed very early on – reportedly by *Hippocrates of Chios* (460-380 BCE) – that finding the side of the cube whose volume is twice that of the cube with side a would require the construction of *two* mean proportionals between a and $2a$. In modern terminology this involves the construction of points x, y such that

$$\frac{a}{x} = \frac{x}{y} = \frac{y}{2a}.$$

Then the cube of the first ratio becomes, after cancellations,

$$\frac{a^3}{x^3} = \left(\frac{a}{x}\right)^3 = \left(\frac{a}{x}\right)\left(\frac{x}{y}\right)\left(\frac{y}{2a}\right) = \frac{a}{2a} = \frac{1}{2}$$

so that $x^3 = 2a^3$. To construct the cube with volume $2a^3$ one needs to construct its side x, and this involves multiplying the length a by the length today described by the number $\sqrt[3]{2}$, the *cube root* of 2. It was only in the

nineteenth century that it became clear – using algebra rather than geometry – why this construction cannot be achieved by straightedge and compass
alone.

Note how the problem posed originally (finding the cube root) has been
subtly transposed into a different one, that of finding two mean proportionals. Rephrasing the original problem in quite different terms is a very common technique in mathematical problem-solving, and often leads to the development of novel concepts that may appear, on the surface, to be quite
unconnected with the original problem. Hippocrates' reformulation (if it is
his!) is one of the earliest examples of this approach.

Conic sections appear in the fourth century BCE, resulting from a search
for two mean proportionals. *Menaechmus* (380-320 BCE) considered curves
arising in this way. Using modern notation, observe that for any constant a,
the first equation in

$$\frac{a}{x} = \frac{x}{y} = \frac{y}{2a}$$

leads to the equation $ay = x^2$, which is the equation of a 'vertical' *parabola*
centred at the origin, while the second becomes $y^2 = 2ax$, which describes
a 'horizontal' *parabola*. Moreover, the identity of the two outer ratios means
that $xy = 2a^2$, which describes a *hyperbola*.

Menaechmus gave two solutions, each of which displayed x, y as the
points of intersection of two of these curves. The first solution used the
hyperbola and the first parabola, while the second employed both parabo-
las. The problem is that conic sections cannot be produced in the plane by
straightedge and compass alone.

2.2. Trisecting the angle. The second 'famous construction problem'
of geometry in antiquity is that of *trisecting* the general angle, that is, to split
it into three equal parts. Just as it is easy to double any square, it is easy to
bisect any angle by straightedge and compass. Draw lines PB, PC to meet
at P. Draw the circle with centre P through B, to cut PC at B'. Let the circle
with centre B, passing through P, meet the circle with centre B' passing
through P at D. Then PD bisects the angle BPC.

Ways to trisect *particular* angles (such as a right angle, for example) were
clearly known to ancient geometers. However, the task of finding a general
method for finding the trisector of *any* angle by straightedge and compass
alone proved to be beyond Greek mathematics, although a number of ingenious curves were invented in their efforts to solve it.

One such curve is attributed to the Sophist *Hippias*, who is said to have
been a contemporary of Hippocrates of Chios. His *quadratrix* cannot be constructed by straightedge and compass, but has to be plotted point by point

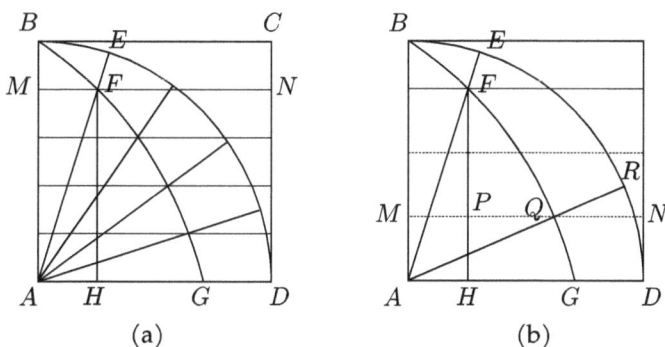

Figure 16. The Quadratrix of Hippias

– hence it was seen as illegitimate by later commentators.[2] It is described by a double motion (see Figure 16(a)):

Describe a quartercircle by rotating the line segment AB clockwise, at constant velocity, through a right angle to AD. At the same time, a line MN moves down from BC to AD at constant velocity, always remaining parallel to both. These constant speeds are chosen so that both reach their final position at the same time. A typical point F on the intersection of these two moving lines (the rotating radius AB and the descending line MN) defines a point on the quadratrix. Since the distance AM decreases, the quadratrix will meet AD in a point G between A and D. Extending AF to E, and drawing in the perpendicular FH to AD, we can describe the quadratrix by comparing the ratios (since the two motions finish at the same time):

$$\frac{\angle BAD}{\angle EAD} = \frac{AB}{FH} = \frac{arcBED}{arcED}.$$

Trisecting the angle EAD (see Figure 16(b)) now amounts to trisecting the line FH (which only needs straightedge and compass – try it!). For if $HP = \frac{1}{3}FH$ and the position of the line MN at this point is shown by $MPQN$, where Q lies on the quadratrix, then the angle QAD is one-third of the angle FAD. To see this, mark the extension of AQ to the circular quadrant by R. Just as above, the ratio of angles is the same as the ratio of vertical lengths, so

$$\frac{\angle EAD}{\angle RAD} = \frac{arcED}{arcRD} = \frac{FH}{PH} = \frac{3}{1}.$$

So angle QAD trisects angle EAD.

---

[2]The quadratrix is so named because it can also be used to solve the third 'famous problem', that constructing the side of a square whose area equals that of a given circle. See **Chapter 8.** Figure 16(b) is used in *MM* for Dinostratus' construction.

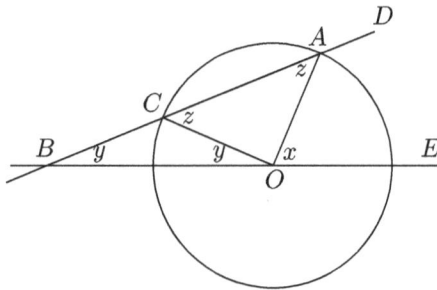

Figure 17. Archimedes' trisection

A very different and rather more straightforward angle-trisector was provided by Archimedes more than two centuries after Hippias. This uses a very natural *neusis* construction, depicted in Figure 17. Suppose that we want to divide angle x, made at O by the lines OD and OE, into three equal parts. Having marked off a length r on our ruler, we can draw a circle of that radius, with centre at O. Extend OE on the opposite side, beyond O, and place the marked ruler at the point A on the circle where it meets the line OD. Then join A with the extension of OE to meet the circle again at C and the extension of OE at B, while ensuring that the distance BC is r. The length requirement $BC = r$ determines the *direction* of line ABC; how it 'inclines towards A' (*neusis.*) To see that angle ABE trisects angle AOE, note that in Figure $z = 2y$ (external angle), while $x + y = 2z$ (on both sides, adding angle AOE produces 180°), so that $x + y = 4y$, hence $x = 3y$.

We can rephrase the angle trisection problem using simple trigonometry. The Greek name for triangle is *trigōnon* – three angles – while *metron* means 'measure', so this subject concerns the measurement of triangles: for example, if we know the lengths of the sides (or, alternatively, sizes of the angles) of a triangle, can we find the angles (or sides) and its area? The key observation from geometry, which we have already used extensively, is that *similar* triangles (that is, with the same shape, since that is determined by their angles) must have their corresponding sides in proportion. In other words, the *ratio* of two such sides does not depend on the size of the triangle, but only on its shape.

The basic ratios of the three sides of a *right-angled triangle* can be described if we make an angle x at the centre A of a circle with unit radius (AC) and complete the right-angled triangle ABC with BC perpendicular to AB (see Figure 18). The ratio $\frac{CB}{AC}$ (the opposite side over the hypotenuse) is then denoted by $\sin x$, the ratio $\frac{AB}{AC}$ (the adjacent side over the hypotenuse) is $\cos x$, and the ratio $\frac{CB}{AB} = (\frac{CB}{AC})/(\frac{AB}{AC})$ is $\tan x = \frac{\sin x}{\cos x}$. Since we chose the

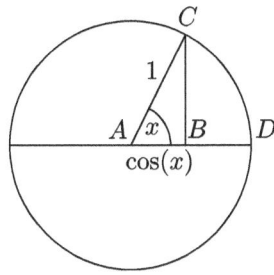

Figure 18. Sine and cosine

length $AC = 1$, Pythagoras' theorem then becomes the identity

$$(\sin x)^2 + (\cos x)^2 = 1.$$

We will follow the usual convention of writing $(\sin x)^2$ as $\sin^2 x$, etc., from now on.

Among the first trigonometric results learned at school are the sum formulae for sine and cosine. For any angles x, y these are:

$$\sin(x + y) = (\sin x)(\cos y) + (\cos x)(\sin y)$$
$$\cos(x + y) = (\cos x)(\cos y) - (\sin x)(\sin y).$$

These formulae can be deduced from a simple geometric proposition, known today as *Ptolemy's Theorem*.[3] This is named after *Claudius Ptolemaeus* (ca. 100-170), author of a famous (geocentric) astronomical treatise, popularly known as the *Almagest*, which held sway until the sixteenth century. The sum formulae enabled Ptolemy to compute extensive *tables* of the values of sine and cosine for different angles, as aids to the extensive calculations needed in astronomical observations.

Beginning with $\cos 90° = 0$, halving the angle provides $\sin 45° = \frac{1}{\sqrt{2}}$ from the above statement of Pythagoras' theorem. More generally, for any angle x we may take $x = y$ in the sum formula for cosine and obtain

$$\cos(2x) = \cos^2 x - \sin^2 x = 1 - 2\sin^2 x,$$

where the final identity again follows from Pythagoras' theorem. Applying this identity to $\frac{x}{2}$ and $x = 2(\frac{x}{2})$ instead, we see that $\cos x = 1 - 2\sin^2(\frac{x}{2})$, so that $\sin(\frac{x}{2}) = \sqrt{\frac{1}{2}(1 - \cos x)}$.

After halving the angle in this fashion several times one arrives at a small angle (eight successive bisections will produce an angle of $\frac{90}{256} \simeq 0.3$ degrees), providing a suitably small gap between successive table entries.

---

[3]See *MM* for details of the statement and proof of this result.

By approximating square roots, Ptolemy could therefore obtain an approximate value for the sine of this small angle, and with this reference value he built tables of sines and cosines for various multiples of the angle, using the above sum formulae. The process involves rather a lot of arithmetic, including the extraction of square roots, but it all quickly becomes routine.

But let us return to angle trisection! Applying the cosine sum formlua to angles $x = 3y = 2y + y$, we easily obtain[4]

$$\cos x = 4 \cos^3 y - 3 \cos y.$$

When x is given, we know the value of $\cos x$ (denoted by the constant c) and the above becomes a cubic equation in the unknown $z = \cos y$, namely $c = 4z^3 - 3z$, which we write in its standard form as

$$4z^3 - 3z - c = 0.$$

If we could solve this cubic equation for z, we would know $\cos y$ and could look up the (approximate) angle y from trigonometric tables. In other words, the problem of trisecting the general angle has been transformed into constructing the solution of a cubic equation, just as we did for the duplication of the cube. In **Chapter 8** we will find out why this construction is impossible when using only a straightedge and compass.

3. Decimals and logarithms

At the end of the sixteenth century, the utility of trigonometric tables was to lead to another crucial aid to calculation, *logarithms*, which quickly became ubiquitous and played a role in the gradual acceptance of irrationals as numbers. This had been made possible by the gradual adoption of the Hindu-Arabic numerals, which greatly facilitated calculation as well as contributing to the acceptance of a wider concept of number, if somewhat haltingly at first. What became our modern decimal notation, including both integral and fractional parts, stems from the late sixteenth century, although, as we have seen, there were earlier examples of similar systems in China, India and especially in Islamic mathematics. However, even the latter was not a fully positional system until the 1500s. We will first trace how our now familiar notation was developed in sixteenth century Europe.

---

[4]

$$\begin{aligned}
\cos x = \cos 3y = \cos(2y + y) &= (\cos 2y)(\cos y) - (\sin 2y)(\sin y) \\
&= (\cos^2 y - \sin^2 y)(\cos y) - 2(\sin^2 y)(\cos y) \\
&= (\cos y)(\cos^2 y - 3(\sin^2 y)) \\
&= (\cos y)(\cos^2 y - 3(1 - \cos^2 y)) \\
&= 4\cos^3 y - 3\cos y.
\end{aligned}$$

3.1. Notation and equations. In France, significant progress was made by *Francois Viète* (1540-1603) who, while trained as a lawyer and serving as privy councillor (and code-breaker) to Henry III and his successor, Henry IV, had a lifelong passion for mathematics and astronomy. He made highly influential contributions to the development of algebraic symbolism and decimal notation. His *Canon mathematicus* (1579) included a systematic approach to the analysis of triangles. It contained new trigonometric identities, as well as trigonometric tables in which he was careful to separate the integral and fractional parts of a number. To show the difference between these he tried various notational devices: first, writing the fractional part in smaller type above a line, then separating the two parts by a vertical line, and finally by showing the integral part in bold type. Thus he wrote an approximation of $100,000\pi$ first as $314,159,\frac{265,36}{100,00}$, later as $314,159|265,36$, and finally as **314**, **159**, $265,36$.

He argued against the use of number systems other than the decimal, such as the sexagesimal system inherited from the Babylonians, while recognising that this had proved useful for large calculations needed in navigation and astronomy. However, Viète stated categorically:

sexagesimals and sixties are to be used sparingly, or never in mathematics, and thousandths and thousands, hundredths and hundreds, tenths and tens, and similar propressions, ascending and descending, are to be used frequently or exclusively.

Viète explored approximations to π that he found in the much admired works of Archimedes, by then translated into Latin from Arabic. He improved Archimedes' estimate using, like his predecessor, the perimeters of regular polygons inscribed in a circle. He also derived what appears to be the first 'infinite product' formula for π, namely

$$\frac{2}{\pi} = \frac{\sqrt{2}}{2} \cdot \frac{\sqrt{2+\sqrt{2}}}{2} \cdot \frac{\sqrt{2+\sqrt{2+\sqrt{2}}}}{2} \cdots$$

where the pattern of the ratios to be multiplied by each other continues indefinitely. This is the first instance where an exact value for π was stated, albeit one involving infinitely many factors in the product! A rigorous proof of his formula was still some way off, however.

Viète's willingness to produce a formula for the exact value of π suggests a significant relaxation of the perception of irrationals that was still prevalent in mid-sixteenth century. An example of this earlier attitude is found in the works of the German monk *Michael Stifel* (1487-1567). Stifel's *Arithmetica integra* (1544) included important innovations such as introducing the term 'exponent', as well as notation and general rules for calculating (integral) powers,

$$x^m x^n = x^{m+n}, \quad \frac{x^m}{x^n} = x^{m-n}$$

and he was among the first to accept *negative numbers* as having equal status to positive numbers. Although Stifel's books do not seem to have been widely read, his position as a professor in Wittenberg helped him to bring aspects of the practical *Coss* tradition (see **Chapter 2**, Section 2) into the formal university curriculum, thus helping to bridge the gap between the two traditions. He formulated rules for arithmetic with negative numbers as well as with fractions, and explored calculating with fractional powers.

Stifel introduced the term *irrational* and accepted the *existence* of irrationals as solutions of geometric problems on the same basis as rationals. However, he did not regard irrationals as 'true numbers', describing them instead as entities *'concealed under a fog of infinity'*. In practice, he nonetheless continued to work with rational and irrational solutions to specific problems, even declaring that there would be infinitely many rationals and infinitely many irrationals between any two whole numbers.

Viète's most famous work, *In artem analyticam isagoge* (1591), introduced much of the symbolic notation for the expression of equations later adopted and adapted by the philosopher and mathematician *René Descartes* (1596-1650). One of Viète's most significant innovations was the use of letters, not only for the unknowns (as the *maestri d'abbaco* and *cossists* had done) but also for the coefficients, the known quantities, in each problem. He distinguished between them by reserving vowels for the unknown, with consonants denoting coefficients. This shifted the emphasis from the origins of a particular problem to the equation itself, as a tool for handling magnitudes of any kind, represented by means of abstract symbols and manipulated according to set rules.

Thus a cubic equation that we now write (in the style of Descartes) as

$$x^3 + cx = d$$

was described by Viète as: *A cubus +C planum in A aequatus D solidum.*

Note that he used words rather than Stifel's much more convenient notation for the exponents. He also continued to insist on dimensional homogeneity, i.e. that the dimensions of the terms in an equation must fit: in our example all three terms represent 'solids' (in the second term, a plane and a line are multiplied – as signalled by *'in'* – to give a rectangular solid). This reflects the geometric origin of his ideas. Nevertheless, he produced *symbolic algebraic rules* for manipulating his various expressions, e.g. for the addition of 'fractional' magnitudes in order to clarify his solution methods.[5]

---

[5]Viete writes ([6], p.157), using *pl* for planum and *in* for multiplication,

$$\frac{Zpl}{G} + \frac{Apl}{B}\ aeq.\ \frac{Gin\,Apl + Bin\,Zpl}{Bin\,G}.$$

In other words $\frac{Z}{G} + \frac{A}{B} = \frac{G \cdot A + B \cdot Z}{B \cdot G}$, which is how we would add these fractions today. Observe how his insistence on dimensional homogeniety complicates the notation.

Viète regarded his methods as *analytic*, in the classical Greek sense, i.e. methods for *finding* solutions to specific problems, thus distinguishing them from assertions made in the *synthetic* Euclidean approach, which were to be legitimated (*proved*) on the basis of a clearly stated system of axioms. In this, he was one of a growing number of Renaissance mathematicians who suspected 'the ancients' of *concealing* their actual solution techniques behind the austere facade of Euclidean geometry. The term *analysis*, introduced by Pappus many centuries before, now came to denote a 'secret method' of finding solutions. René Descartes later voiced this suspicion most clearly (see [6], p.159ff) in his *Rules for the Direction of the Mind*:

I was confirmed in my suspicion that they had knowledge of a species of mathematics very different from that which passes current in our time...I seem to recognise certain traces of this true Mathematics in Pappus and Diophantus...

But my opinion is that these writers then with a sort of low cunning, deplorable indeed, suppressed this knowledge...

... Finally, there have been certain men of talent who in the present age have tried to revive this same art. For it seems to be precisely that science known by the barbarous name of Algebra, if only we could extricate it from that vast array of numbers and inexplicable figures by which it is overwhelmed, might display the clearness and simplicity which we imagine ought to exist in a genuine Mathematics.

It seems clear that Descartes has Viète in mind in this passage as one of the 'men of talent' who had revived the science of Algebra (so named by 'barbarian' Arabs!). This highlights the importance of the conceptual steps taken in Viète's work, giving priority to the algebraic formulation of problems and describing new algebraic techniques for their solution, thus preparing the ground for the *analytic geometry* of Descartes and Fermat we describe in **Chapter 4**.

But, quite apart from theoretical efforts to rediscover the 'lost analysis of the ancients', more practical concerns to expedite calculations needed in navigation, astronomy, commerce and engineering, and to make them more widely accessible, were to dominate the search for a unified *arithmetical notation* that would describe both discrete multiples and continuous magnitudes.

In 1585 the Flemish engineer *Simon Stevin* (1548-1620) published a pamphlet *De Thiende* (The Tenth), quickly translated into French and widely distributed, which introduced a consistent notation to describe fractions in terms of negative powers of 10 (tenths, hundredths, thousandths, etc.). Much as we think of minutes and seconds as integers rather than as fractions of an hour, he avoided Viète's denominators altogether, writing the familiar approximation of π to four decimal places as $3(0)1(1)4(2)1(3)6(4)$, where the numbers in brackets show the power of 10 by which the previous digit should be divided. (Stevin actually used circles rather than brackets.) In current notation this is 3.1416. Anticipating ideas developed more fully by *Karl-Weierstrass*and others in the late nineteenth century, he also considered irrationals to be determined by a *nested sequence* of finite decimal fractions: he starts with approximations that, respectively, under- and over-estimate the irrational in question. The distance between such under- and over-estimates decreases indefinitely as ever better estimates are found. Stevin argued that in this fashion 'one approaches the desired value infinitely closely'.

He was explicit in his view that the ancient distinction between multiples and magnitudes was unhelpful and should be abandoned. In another text published in 1585 (*L'arithmétique*) he followed through on his view, insisting, among other things, that 1 should be seen as a number (although he still did not accord the same status to 0, even though he used the symbol freely and regarded it as *'le vrai et naturel commencement'* of numbers, just as a point may be seen as the beginning of a line). He also argued that there are *'no absurd, irrational, irregular, inexplicable or surd numbers'*, thus placing all numbers on the 'number line' on the same footing. While he accepts negative numbers, he does *not* follow Bombelli into the realm of the 'imaginary', to what we now call complex numbers: *'There are enough legitimate things to work on without need to get busy on uncertain matter'*, he argued.

Although Stevin's decimal notation was still cumbersome for calculation, his clear explanation of the principles underlying decimal fractions made *De Thiende* very popular. The Scottish landowner *John Napier* (1550-1617), now chiefly celebrated for the tables of logarithms he published in 1614, suggested that Stevin's notation could be improved by the use of a decimal point or comma. This was quickly taken up. Although the *decimal point*, as shown in our 3.1416, had first been used by two associates of *Johannes Kepler* in the 1590s, its use became standard in Britain only from 1619, following the posthumous publication of Napier's *Construction of the Wonderful Canon of Logarithms*, which explained the theoretical underpinnings of his tables. In much of the rest of Europe, however, the *comma* became widely used in the same role—perhaps in deference to Viète?—and largely remains so today!

3.2. Napierian logarithms. The ideas that led to logarithms were clearly in the air by 1600. Tables relating powers of 2 to their exponents and showing that multiplication led to addition of the powers had featured in earlier work of Stifel and Chuquet. The Swiss clockmaker *Jobst Bürgi* (1552-1632) had independently constructed tables of logarithms, perhaps as early as 1588, but Napier's were the first to be published, in his *Mirifici logarithmorum canonis descriptio (Description of the Wonderful Canon of Logarithms)* in 1614.

Napier's system of logarithms was developed principally as a computational aid, to remove the drudgery from long multiplication and division in navigation and astronomy. His basic idea was to use *geometric progressions* of successive powers of a fixed number; that is, $1, x, x^2, x^3, ..., x^n, ...$ since then the exponent $m+n$ of the product $x^m x^n = x^{m+n}$ is the sum of the exponents of the two numbers being multiplied. In order to produce workable tables, the initial number should be close to 1; otherwise successive powers of that number would grow apart too quickly.

Napier decided to use the number $1 - 10^{-7} = 0.9999999$ as his reference point. Rather than deal with decimal fractions, he then multiplied integral powers of this number by 10^7. Thus the *'logarithm'* (a term Napier coined from the Greek words *logos* (logic) and *arithmos* (number) respectively) of the number $10^7(1 - 10^{-7})^L$ is given by the exponent L. Therefore $L = 1$ is the logarithm of $10^7(1 - 10^{-7}) = 9,999,999$ and $L = 0$ is the logarithm of $10^7(1 - 10^{-7})^0 = 10^7$.

Napier imagined the number 10^7 as the hypotenuse of a very large right-angled triangle ('whole sine') that provided the starting point for his calculations.[6]

He describes the relation between the geometric progression of successive powers of $(1 - 10^{-7})$ and the arithmetic progression of the associated exponents by reference to motion: *'The logarithme, therefore, of any sine is a number very neerely expressing the line which increased equally in the meene time whiles the line of the whole sine decreased proportionally into that sine, both motions being equal timed and the beginning equally swift.'*

We can interpret this by imagining a point moving along a fixed line of length 10^7 units. Its velocity at each instance is inversely proportional to the distance left to travel. Thus, at time 0, the distance remaining is 10^7 units, at time 1 it is $10^7(1 - 10^{-7})$, at time 2 the remaining distance is $10^7(1 - 10^{-7})^2$, and so forth. The uniform 'second motion' is then represented by the passage of time, i.e. $0, 1, 2, ...$ units, until we reach 10^7. The two sequences of numbers are arranged in a two-row table, with the time points forming an arithmetic progression (here, simply adding 1 at each time) in the first row, and the corresponding positions of the 'moving point' forming the second row, as a (decreasing) geometric progression.

In terms of modern logarithms we would regard $9,999,999$ as the *base* of the Napierian table, since for Napier the logarithm is 1; however, this concept was not used by Napier. Dividing each term by 10^7, his geometric progression has $x = 0.999999 < 1$, so his arithmetic progression of logarithms increases while geometric progression decreases. Moreover, his logarithms do not satisfy the basic identity that 'the logarithm of the product is the sum

---

[6]In Napier's time the 'sine' was seen as a line segment, not as a ratio. For the unit circle in Figure 18 the line CB (representing what we called the sine) is a half-chord of the circle centred at A. Similarly, what Napier calls the 'whole sine' is the radius AC of a circle whose half-chord is the 'sine' CB. Napier spoke of the logarithm of a sine; his principal objective was to simplify trigonometric calculations.

of the logarithms':

$$\log(xy) = \log x + \log y.$$

This is the formula subsequent generations learnt in school over the next 400 years. To see how Napier's construction differs from that used later, we write the above logarithms of Napier with a capital L. Now $L_1 = Log(x)$ means that $x = 10^7(1 - 10^{-7})^{L_1}$ and similarly for $L_2 = Log(y)$. Then $xy = 10^{14}(1 - 10^{-7})^{L_1+L_2}$ and we must now *divide* by 10^7 to obtain $\frac{xy}{10^7} = 10^7(1 - 10^{-7})^{L_1+L_2}$ as the number whose logarithm is $L_1 + L_2$. In other words, Napier's logarithms follow the rule that

$$Log(x) + Log(y) = Log(\frac{xy}{10^7}).$$

Thus using the tables to look up the 'antilogarithm' of the sum on the left yields (for Napier) the antilogarithm of $\frac{xy}{10^7}$ instead of that of xy. Although Napier's logarithms were a great improvement on earlier methods and were instantly recognised as such, it was clear that there remained room for simplification.

3.3. Briggs' logarithmic tables. As Napier was nearing the end of his life, this task was undertaken by the first Savilian professor of geometry in Oxford, *Henry Briggs* (1561-1630), who visited Napier in 1615. They agreed that using powers of 10 would be preferable, and decided that Briggs should create a system where $\log_{10} 1 = 0$ and $\log_{10} 10 = 1$ (since $10^0 = 1$ and $10^1 = 10$).

These became the 'logarithms to base 10' tables that were routinely used for calculation in schools until perhaps forty years ago. The once ubiquitous 'log books', containing tables of logarithms of numbers and of trigonometric ratios, have now been superseded by electronic calculators, which school pupils routinely treat as 'black boxes' that magically produce answers to complicated calculations, without divulging the method by which they were found! While the practical benefits of calculators are clear, it remains to be seen if anything of conceptual or educational value has been lost in this change.

Briggs' first tables included logarithms of whole numbers up to $1,000$ as well as logarithms of sines, which were of particular use in astronomical calculation. Both tables were expanded significantly in his *Arithmetica Logarithmica* (1624) which included logarithms of whole numbers from 1 to $20,000$ and from $90,000$ to $100,000$, all to 14 decimal places! The remaining gaps were filled in by his Belgian publisher Adriaan Vlacq and Ezechiel de Dekker for the second edition, published in 1629.

The burden of calculation involved in finding these logarithms was considerable, but the impact of the tables in assisting large calculations in many practical contexts was immediate and the use of logarithmic tables became widespread very rapidly, saving much tedious effort for the next four centuries.

In fact, most logarithms are themselves *irrational* numbers. To see that, for example, the common (i.e. Briggsian) logarithm of 2 cannot be rational, consider the following simple argument. Let us suppose that $\log_{10} 2 = \frac{p}{q}$ for some whole numbers p, q. This would entail that $10^{\frac{p}{q}} = 2$, so that $10^p = 2^q$. But powers of 10 all have their final digit as 0, while powers of 2 end in 2, 4, 6 or 8. This contradiction shows that $\log_{10} 2$ cannot be of the form $\frac{p}{q}$, so it is not a rational number.

In the sixteenth century irrationals could not really have been regarded as well understood. However, this fact seems not to have deterred anyone from recognising logarithms as a spectacularly successful aid to calculation. Approximation to a fairly small number of decimal places was quite sufficient for all practical applications, and the utility of the new calculation techniques allowed practitioners to suppress any concerns about the precise nature of the numbers that were being approximated. Their success further blurred the Ancient Greek distinction between the discrete and the continuous, and thus contributed to a broader and more abstract approach to the concept of number. These developments proved to be a vital precursor to the conceptual revolution begun in the 1630s by René Descartes and Pierre de Fermat, which is discussed next.

CHAPTER 4

Coordinates and Complex Numbers

If a man will begin with certainties, he shall end in doubts, but if he will be content to begin with doubts, he shall end in certainties.

Sir Francis Bacon, *The Advancement of Learning*, 1605

Summary

The next crucial steps that took mathematicians towards a fuller understanding of the structure of polynomial equations—leading, almost incidentally, to a more inclusive approach to the concept of number—were taken, apparently independently, by two Frenchmen: the lawyer *Pierre de Fermat* (1601-1665) and the philosopher *René Descartes* (1596 -1650). Both took initial steps toward the introduction of what we call a *coordinate system*, a term coined somewhat later by the German philosopher *Gottfried Wilhelm Leibniz* (1646-1718). This enabled them systematically to reformulate, solve and generalise many geometric problems inherited from Ancient Greece by using the algebraic formalism developed by Arabian mathematicians and their Renaissance counterparts in Europe.

This chapter highlights the innovations in Descartes' revolutionary contribution and to widespread use of algebraic notation, paving the way for acceptance of the system of *real numbers* (the 'number line') and [1] to what is now known as analytic geometry. His work inspired his successors, including Isaac Newton, to remove suspicions about negative numbers, and to a significant extent, to accept irrationals as numbers. However, the nature of square roots of negative numbers was clarified only in the nineteenth century, principally through the influence of Gauss and by Hamilton's abstract definition of the complex number system. This led to what became known as the *Fundamental Theorem of Algebra*.

---

[1]Our focus will be on Descartes because Fermat did not publish his short treatise *Introduction to Plane and Solid Loci* in his lifetime. After his death it was published by his son in 1679, by which time Descartes' methods had already been widely disseminated and developed further by others.

https://doi.org/10.11647/OBP.0236.04

1. Descartes' analytic geometry

1.1. *Discours de la Méthode*. René Descartes is known today principally as a philosopher: his *Discours* begins with an account of his rigorous search for indubitable truth, which led him to his dictum *'I think, therefore I am'* ('Je pense, donc je suis' – later more widely known in its Latin translation *cogito ergo sum*). His reasoning was essentially that doubt, as an act of thought, cannot occur unless there is a thinker. He continued his search for further propositions that he should accept as certain, arguing that *'we ought never to allow ourselves to be persuaded of the truth of anything unless on the evidence of our reason.'*

Descartes' principal mathematical text appeared in 1637 as his (now famous) *La géométrie*, published as one of three annexes to his highly influential *Discours de la Méthode*. I will focus on how Descartes' new methodology led to the *coordinate geometry* nowadays routinely taught at school level – a development which the English philosopher *John Stuart Mill* (1806-1873) later described as *'the greatest single step ever made in the progess of the exact sciences'*.

Descartes' methodology differs fundamentally from what we now regard as *scientific method*, since he does not accept that our understanding of natural phenomena should be based on empirical observation. Rather like Plato, he wishes it to be based on metaphysical principles and on 'basic facts' of physical reality which in turn should be derived from the 'indubitable truths' supplied by 'pure reason'. In a letter to his friend, the priest *Marin Mersenne*, he chides Galileo, who *'without having considered the first causes of nature, has merely looked for the explanations of a few particular effects, and he has thereby built without foundations'*.

1.2. Coordinate systems. In Descartes' mathematical researches, however, the foundations were already in place, due largely to the work of Viète (see **Chapter 3**, Section 3.1). In *La géométrie*, his main concern was to bring order to geometric constructions (*'the analysis of the Greeks'*) by combining them with modern algebraic notation (*'the algebra of the moderns'*). The synthesis he proposed would free geometry from its reliance on diagrams and, at the same time, give concrete meaning to algebraic operations through their geometric interpretation.

Descartes' goal remains a geometric construction, but his method of achieving this makes full use of algebraic operations. The crucial link between the two is created by choosing a point O in the plane, and a pair of reference lines that meet in O. Today, the reference lines are called *axes*, typically labelled as X and Y respectively. The position of any point P in the plane can then be described algebraically by means of a pair (x, y) of *numbers* that measure the distances needed to get from O to P by moving only in the directions of the two axes: from O we first move a distance x along

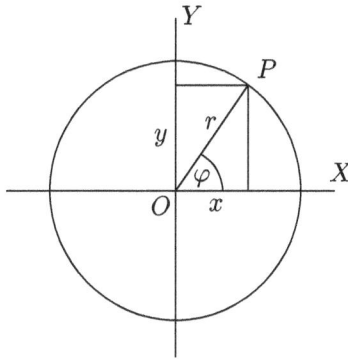

Figure 19. Rectangular and polar coordinates

the X-axis, and, from there, a distance y on a line parallel to the Y-axis, to reach P (or, alternatively, move distances y along the Y-axis and x on a line parallel to the X-axis). The path traced out by a curve or a geometric figure can then be described via an equation that relates these x and y coordinates by setting out in algebraic terms the geometric conditions that any point $P = (x, y)$ on the curve (or figure) must satisfy.

Today we regard Descartes and Fermat as the originators of *analytic* (or *coordinate*) *geometry*, in which we choose two *perpendicular* axes X and Y which, as shown in Figure 19, define the *Cartesian plane* (named in honour of Descartes). Any point P in this plane is then described by the distances (x, y) from O to the *projections* of P onto the (X, Y)-axes. This definition has the advantage that the *length* $\sqrt{x^2 + y^2}$ of the line segment OP is given immediately by Pythagoras' theorem: for example, the collection of points $P = (x, y)$ at distance r from O is (by definition) the circle with centre O and radius r, and the points x, y are related by $x^2 + y^2 = r^2$, as the triangle in Figure 19 shows.

However, neither Descartes nor Fermat actually used such a rectangular coordinate system. In fact, as we will indicate below, Descartes' system involved positive coordinates: in our terms, when using rectangular coordinates, we would call this the *positive quadrant*. The development of a full coordinate system was undertaken by his successors. Nevertheless, the key breakthrough – linking the apparently unrelated areas of geometry and algebra – had been made, making simplifications such as rectangular coordinates and extensions to more than two dimensions much easier to achieve.

Descartes further elaborated the algebraic symbolism he had inherited from Viète, and his algebraic expressions are essentially those in use today. Crucially, he also dispensed with the need to give all terms in a polynomial (such as a cubic or quartic) equation the same 'dimension', as Omar

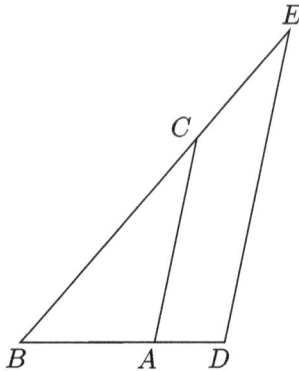

Figure 20. Descartes' product: $BA = 1, BE = BC.BD$

Khyayam had done. Unlike modern usage, and more in keeping with Greek tradition, he regarded the known parameters and the unknowns in his equations as lengths of lines, rather than as numbers. Yet he broke with Greek usage by considering x^2, x^3 as *lines*, not as areas or volumes. For example, instead of treating the product of two line segments as a rectangle, he defined their product by using similarity of figures and proportionals.

It is the systematic introduction of a *unit length* in any geometric problem that provides the key to Descartes' approach. His geometric construction of the product ab of two line segments was extremely simple, as shown in Figure 20: letting two given line segments BC and BD meet at B (at an arbitrary angle), he chose an arbitrary distance BA along BD as the unit (extending the segment if necessary) and drew the line segment AC. Then he drew the line DE parallel to AC, to meet BC (extended) at E. The triangles ABC and DBE are obviously similar, which ensures that the ratios of corresponding sides are in proportion. Hence $BC : BA :: BE : BD$. Choosing BA as the unit length, while $BC = a$ and $BD = b$, this shows that BE becomes the product $BC.BD = ab$. Using the same figure (and unit) the other way round one obtains BC as the quotient of BE by BD. Both product and quotient are therefore realised as line segments.

The construction of a mean proportional (**Chapter 3**, Figure 15) demonstrates how Descartes could interpret the square root similarly as a line segment: to find the square root of a line segment OA, extend it by a chosen unit AB to OAB with $AB = 1$, and from its midpoint draw a circle through O. The perpendicular AC from its circumference provides the desired square root of OA .

Descartes' notation when describing geometric problems was essentially algebraic. He used and extended Viète's new algebraic symbolism: *'it is sufficent to designate each* [line segment] *by a single letter. Thus, to add the lines BD*

and GH, I call one a and the other b, and write a + b. Then a − b will indicate that b is subtracted from a; ab that a is multiplied by b; a/b that a is divided by b; ...'.

As he points out, early in *La géométrie*, dimensional homogeneity in an equation can always be restored by multiplying or dividing a term sufficiently often by the chosen unit! For example: *'if it be required to extract the cube root of $a^2b^2 − b$, we must consider the quantity a^2b^2 divided once by the unit, and the quantity b multiplied twice by the unit.'* (See [**6**].)

Today this explanation may appear trivial and unnecessary, as the symbols are regarded as *abstract entities*, with no need to think of them in terms of geometric dimensions. Yet it is precisely this essential insight by Descartes that eased the way for his successors to regard the constants, coefficients and unknowns occurring in equations as symbols that can be represented by *numbers*. As *Fauvel and Gray* observe in [**12**], the simple step of eliminating dimensional considerations *'...turns out to have lifted a weight off everyone's shoulders...'*

Descartes systematically used the initial letters of the alphabet for known quantities, leaving final letters such as x, y, z to denote the unknowns. He applied the exponential notation, by then commonly used, to the unknowns (although, while happy to write x^3, x^4, etc., he persisted in writing the square as xx rather than x^2), and he used the Germanic symbols $+, −$ for for addition and subtraction. He did not use Robert Recorde's symbol $=$ for equality; the symbol he employed, possibly drawn from the abreviation 'æ' of the Latin 'aequalis', remained popular on the Continent for some time.

1.3. La géométrie. The first two books of *La géométrie* focus on applying algebraic techniques to geometry. Giving the 'unknown' a *name*, despite not knowing its value in advance, was an important step forward. It enabled him to treat known and unknown quantities on the same basis. His approach was to attack a geometric problem by converting it into an algebraic equation, simplifying the equation as far as possible and then to solve the equation geometrically. Given any geometric problem, his method for its solution is the following (the quotations are from [**12**]):

'...we first suppose the solution already effected, and give names to all the lines that seem needful for its construction—to those that are unknown as well as those that are known. Then, making no distinction between known and unknown lines, we must unravel the difficulty in any way that shows most naturally the relations between these lines, until we find it possible to express a single quantity in two ways. This will constitute an equation, since the terms of one of these two expressions are together equal to the terms of the other.'

Descartes gives a simple example to show how an equation obtained in this way, once simplified as far as possible, can be solved geometrically. Figure 21 illustrates his argument.

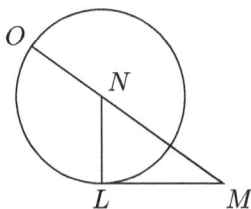

Figure 21. Example from *La Geometrie*

'*...if I have $z^2 = az + b^2$, I construct a right triangle NLM with one side LM, equal to b, the square root of the quantity b^2, and the other, LN, equal to $\frac{1}{2}a$, that is, to half the other known quantity which was multiplied by z, which I supposed to be the unknown line. Then, prolonging MN, the hypotenuse of this triangle, to O, so that NO is equal to NL, the whole line OM is the required line z. This is expressed in the following way:*

$$z = \frac{1}{2}a + \sqrt{\frac{1}{4}a^2 + b^2}.$$

He recognised that the *degree* of the algebraic equation expressing the curve in terms of the (X, Y)-coordinates would determine the means by which the geometric construction of its roots can be achieved:

'*If it can be solved by ordinary geometry, that is by the use of straight lines and circles...there will remain at most only the square of an unknown quantity...*'

In other words, straightedge-and-compass constructions lead to linear or quadratic equations, and he showed how to solve these. Descartes knew that Viète had shown the duplication of the cube and angle trisection to lead to cubic equations, and he now asserted that these constructions could not be effected by straightedge and compass alone – although his attempt at a proof was defective. He argued, as had the Greek geometers, that all constructions should be effected by the simplest means possible; for him, this was determined by the degree of the equation. Quadratics could be handled by straightedge-and-compass, cubics and quartics required conic sections.

Descartes was encouraged by his success in dealing with a set of problems posed by the Alexandrian geometer *Pappus* (fourth century) which attracted much attention at the time. The simplest, the so-called *three-line locus problem,* assumed that one was given three lines in the plane. Pappus sought the *locus* of points such that the product of their distances from two of these lines is proportional to the square of the distance from the third

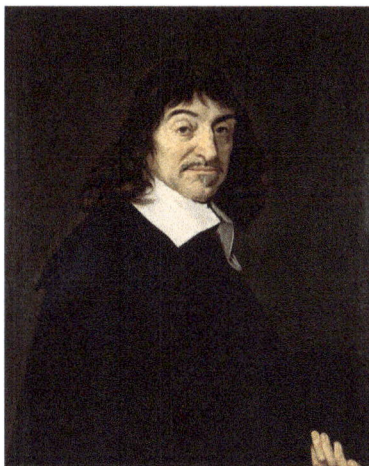

Figure 22. René Descartes by Frans Hals, 1649[3]

line.[2] Descartes chose one of the three lines as AB and fixed the position of a typical point C of the locus he sought by marking AB as x and BC as y, so that these two became the reference lines (what we would call the *axes* today) for his system of coordinates. Expressing the other lines in terms of these two quantities he arrived at an equation (involving the variables x and y) that the point C must satisfy. He solved the problem by simplifying this equation as far as possible and effecting a geometric construction for the roots of the simplified equation.

In the second book he continued to extend this approach to a larger set of lines than Pappus had done and suggested a general classification of geometric problems: those leading to quadratic equations formed the first class and could be solved by straightedge and compass constructions, those leading to cubic or quartic equations he placed in the second class, since the roots could be constructed using conic sections. Quite generally, a problem of class n was associated with an equation of degree either $2n - 1$ or $2n$. He conjectured – incorrectly, as it turned out – that, since the solution of the quartic can be reduced to that of an associated cubic (as Ferrari had shown), a similar argument could always be found to solve an equation of degree $2n$ by reducing its solution to that of an equation of degree $2n-1$. Although his classification did not hold water, his work encouraged the much freer use of higher plane curves in geometric constructions. A famous example is the

---

[2]A locus is the collection of points satisfying some given geometric condition(s). For example, a circle is the locus of points lying at a fixed distance from a given point (its centre), while an ellipse is the locus of points for which the sum of their distances from two given points (its foci) remains constant. Thus a circle is an ellipse whose foci coincide.

[3]https://commons.wikimedia.org/wiki/File:Frans_Hals_-_Portret_van_René_Descartes.jpg

folium of Descartes, which makes a loop in the first (i.e. 'north-east') quadrant, intersecting itself at the origin, while its two ends go off to infinity in the second ('north-west') and fourth ('south-east') quadrant of our rectangular (X, Y)-plane. Its equation has the form $x^3 + y^3 - 3axy = 0$.

He went on to classify as *'geometric curves'* all those whose points can be determined by the intersection of two lines, each moving parallel to one of his axes (X or Y) with *commensurable* velocities (whose ratio takes the form $\frac{m}{n}$ for some whole numbers m, n). His study of Pappus' three- and four-line problems led him to the general equation of a conic passing through the origin in the form: $y^2 = ay - bxy + cx - dx^2$, where he identified the types that can occur for different choices of the constants a, b, c, d.

Modern terminology, following a more general classification introduced by Leibniz half a century later, defines an *algebraic curve* as the collection of points (x, y) in the plane satisfying a given polynomial equation in the variables x, y. In particular, the points (x, y) on a straight line satisfy the first-degree equation

$$ax + by + c = 0,$$

while the general quadratic equation in x, y,

$$ax^2 + bxy + cy^2 + dx + ey + f = 0$$

will give rise to a conic section (see Figure 14 in **Chapter 3**). The simplest examples of conic sections occur when we have rectangular axes and centre the figure at the origin and symmetric to the axes. The circle of radius r is then given by the points (x, y) satisfying $x^2 + y^2 = r^2$, an ellipse will have the form $(\frac{x}{a})^2 + (\frac{y}{b})^2 = 1$, a hyperbola is given by $(\frac{x}{a})^2 - (\frac{y}{b})^2 = 1$, a parabola typically by $y^2 = 4ax$.

Similarly, equations involving higher powers of the variables x, y have their *degree* specified by the highest power of these variables – so that the folium of Descartes has degree 3, for example. Leibniz called curves that cannot be specified by such a polynomial equation *transcendental*.

In his (earlier) classification Descartes in effect regarded as 'geometric' all curves we now describe as 'algebraic'. His primary objective was to establish a criterion that could be expressed in *geometric* terms, rather than our algebraic one, to describe such curves. His definitions implicitly assume that all geometric curves can be traced by a *continuous motion*. Thus the quadratrix, and other curves generally defined in terms of arc lengths, were to be excluded from his classification of geometric curves, since they arise from two simultaneous motions *'whose relation does not admit of precise determination'*. For the quadratrix, one motion is a translation, the other a rotation, and, as we saw in Figure 16(a), the ratio of the lengths of the arc BED and the radius BA is $\frac{\pi}{2}$, which is irrational; in other words, the two lines are incommensurable. Descarted stated this claim without proof – in effect, he claimed that the circle cannot be squared! He argued that such curves should be excluded from geometry:

Geometry should not include lines (or curves) that are like strings, in that they are sometimes straight and sometimes curved, since the ratios between straight and curved lines are not known...

Consequently, Descartes called such curves *mechanical*.

In Figure 19 we also depict an alternative description of coordinates, first introduced by the Italian *Bonaventura Cavalieri* (1598-1647) in order to discuss the 'Archimedean spiral', which will feature in **Chapter 5**. This system of *polar coordinates* became widely used after *Jacob Bernoulli* (1654-1705) employed it more systematically. One chooses an origin O, or *pole*, as well as a directed line segment OX, which serves as the *polar axis*. The length OP then describes the *radial distance* r of a point P in the plane from the pole, and the angle XOP (denoted by ϕ, and taken anti-clockwise) determines the direction of OP relative to the chosen direction of the polar axis. Pythagoras' theorem and simple trigonometry show that the polar and rectangular coordinate systems are related by:

$$r = \sqrt{x^2 + y^2}, \tan \phi = \frac{y}{x}$$
$$x = r \cos \phi, y = r \sin \phi.$$

Despite his dismissal of 'mechanical' curves from geometry, in 1638 Descartes was led (when considering the path of an object falling towards a rotating Earth) to a 'mechanical' curve that contradicted his assumption that no such curve could be *rectified* (this term means that we can construct a straight line whose length equals that of the curve). The curve in question is the *logarithmic spiral*, which is most simply defined in polar coordinates by $r = ae^{b\phi}$, where a, b are constants, and e is the base of 'natural' logarithms (which we will meet in **Chapter 5**). The Italian mathematican *Torricelli* (1608-1647) was the first to rectify this curve in 1645, by methods that foreshadowed Newton and Leibniz' invention of the Calculus over 20 years later.

2. Paving the way

The 'marriage' of geometry and algebra by Descartes and Fermat served to accelerate the acceptance of irrationals as genuine numbers. Descartes' classification of geometric curves included many different types of irrational roots which could be defined as distances from the origin O along the X-axis, in exactly the same way as rational roots. In handling these quantities algebraically there was no need to distinguish between different sorts of lengths if one simply regarded them as points on a *number line*.

2.1. The number line. The coordinate system provides a visual representation, not only of the curve being analysed, but also of solutions of the equation defining the curve. These appear as *points* on the X-axis, whether positive or negative, and can be treated as *numbers*, whether rational or irrational. As this provided a convenient visual (and *practical*) way of avoiding the inevitable philosophical question of what irrationals actually *are*, the more difficult problem of finding a viable *arithmetical definition* of irrationals was essentially shelved until the nineteenth century, by which time the previously unchallenged centrality of Euclidean geometry had become a serious issue for debate.

Throughout the eighteenth century, mathematicians were generally much too busy exploiting the rich rewards of the new techniques offered by Descartes' analytic geometry, and the competing formulations of the Calculus by Newton and Leibniz a few decades later, to concern themselves in detail with this philosophical issue. They seeemed content to regard as *real numbers* any points on the number line (positive or negative, rational or irrational).

Today it has become a commonplace to conceive of this 'number line', centred on some point O and extending indefinitely to left and right, as representing the *real number system* $\mathbb{R}$ upon which most common modern mathematical structures rest. For practical measurements, of course, we must always content ourselves with rational approximations (such as 3.14159 for π, 2.71828 for e, or 1.4142 for $\sqrt{2}$) since our mechanical or electronic instruments all have physical limitations. But today we nevertheless endow irrational numbers with as much 'reality' as we do 5 or $\frac{94}{73}$. This applies equally whether they are positive or negative, since negative numbers, represented by points lying to the left of the origin O on the number line, are regarded as just as 'real' as positive numbers.

This perspective, due in good measure to the ubiquity of coordinate systems, lends weight to the argument that numbers may be seen as *abstract* entities. My personal preference (not shared universally among mathematicians, as we shall see later) is to regard them as *human inventions* that assist us collectively in making meaningful assertions about the world around us, rather than being or representing actual objects that 'exist' independently of us. This viewpoint differs markedly from the Platonic perception that numbers (and lines), as idealised abstractions, exist in some unseen 'World of Ideas'.

Descartes' analytic geometry may have had a significant impact on the gradual acceptance of a wider, more abstract, concept of number. Historically, however, it took the resolution of the further puzzle of finding a satisfactory visual representation of *'imaginary roots'* (roots like $\sqrt{-1}$, which occur in equations like $x^2 + 1 = 0$, and were given this name by Descartes) before our modern perspectives were fully accepted. These developments are outlined in the next section. Meanwhile, the nomenclature used even to-day (using the term *real numbers* for elements of the number line, while $\sqrt{-1}$ is often called the *imaginary unit*), reflects this tortuous history and retains its potential for misconceptions.

Like Napier's invention of logarithms, Descartes' analytic geometry techniques, combining algebra and geometry, found acclaim upon their publication in 1637, and were developed and refined further by his contemporaries. This process involved considerable effort. Despite Descartes' insistence that one should always begin with the simplest possible constructions before moving on to more complex ones, his discussion of various geometric problems in *La géométrie* was hardly systematic, focussing instead on specific, often rather difficult, problems.

He also had a habit of leaving many details of his verifications to the reader. A frequent refrain in the text was *'it already wearies me to write so much about it'*. He justified his often sketchy solutions by saying that he had omitted details *'in order to give others the pleasure'* of discovering things for themselves! This exacerbated the difficulties that his initial audience found in understanding his methods, and gave impetus to extensive commentaries by other mathematicians. Prominent among these was a Latin edition of *La géométrie* by the Dutch mathematician *Frans van Schooten* (1615-1660), which appeared in 1649 and had seen four editions by 1700, firmly cementing Descartes' analytic geometry in the Continental mathematical tradition.

2.2. Wallis and Newton on numbers. In Britain, the algebraic approach pioneered by Descartes was enthusiastically taken up by *John Wallis* (1616-1703), who held the Savilian Chair of Geometry at Oxford University from 1649. Ordained as a priest, he had been active in decoding Royalist messages for the Parliamentary side in the English Civil Wars, and had studied earlier mathematical works by William Oughtred and Thomas Harriot which had introduced him to the new methodologies developed on the Continent.

A highly original mathematician, he published *Arithmetica Infinitorum* in 1656, tackling many then prevalent problems—area and volume calculations, and finding tangents to various curves—by pioneering mainly arithmetical methods involving *infinite* sums and products. Prominent among these was his infinite product formula for π.[4] This allowed him to approximate π (the symbol $\prod_{n=1}^{\infty}$ means successive multiplication, $n = 1, 2, 3, ...$):

$$\frac{\pi}{2} = \prod_{n=1}^{\infty} \left(\frac{2n}{2n-1} \cdot \frac{2n}{2n+1} \right) = \left(\frac{2}{1} \cdot \frac{2}{3} \right) \cdot \left(\frac{4}{3} \cdot \frac{4}{5} \right) \cdot \left(\frac{6}{5} \cdot \frac{6}{7} \right) \cdot ...$$

---

[4]See MM for details.

Wallis treated a proportion simply as asserting the equality of two fractions, awarding ratios the same status as whole numbers. Although this marked a departure from long-held views, it seemed entirely natural to him. He remained less certain about the status of negative numbers, since in his view it was impossible for a quantity to be *'Less than Nothing, or any number fewer than None'* [6]. But he recognised the usefulness of accepting negative numbers in calculations. Using analogy with movement to the right or left of a starting point, he argued that negative numbers could be represented by points to the left of a chosen origin on a number line. He went further, seeking to represent *square roots* of negative numbers geometrically, using a construction similar to that of the mean proportional (Figure 15).[5] He argued that, while their status as numbers was uncertain, using them in calculations was *'not altogether absurd'*. Wallis also maintained doubts about the status of irrationals as numbers, but nonetheless used them freely in his calculations, arguing, as Stevin had done, that they can be approximated arbitrarily closely by fractions.

Wallis' views were not universally accepted by English mathematicians. At Cambridge, Newton's mentor *Isaac Barrow* (1630-1677) was prominent in something of a backlash against algebraic methods for the solution of geometric problems, although even he used symbolic representations and numerical examples in his widely read 1655 edition of Euclid's *Elements*. He criticised Wallis' assertion that arithmetical equalities had a meaning independent of geometric interpretation, and argued that irrationals like $\sqrt{2}$ were best understood in terms of geometric ratios, rather than in terms of numbers and fractions. For him classical geometry, based upon axioms, provided a clarity of meaning that algebra had not yet achieved.

Isaac Newton (1642-1727) succeeded Barrow to the Lucasian Chair of Mathematics at Cambridge in 1669. An account of Newton's views on numbers can be found in his *Arithmetica Universalis*, first published in 1707 (an English translation, *Universal Arithmetick*, by Joseph Raphson—to which Newton refused to add his name—followed in 1720). This work is not, however, among his best-known today. The focus in the book is on the practice, rather than foundations, of the new algebraic techniques, and features many illustrative examples. Perhaps reflecting Barrow's influence, Newton stresses his preference for classical geometry in many of his comments in the text.

The origins of the text lie in drafts and lecture notes dating from the period between 1673 and 1683, when he had studied Cartesian methods closely and critically. The volume is not a carefully edited and polished publication in the spirit of his 1687 *Principia Mathematica*. Newton was only persuaded to agree, reluctantly, to the publication of *Arithmetica Universalis* (which was not overseen by him, but by his successor to the Lucasian Chair, William Whiston) when he needed to attract financial support from his academic

---

[5]See e.g. [6] for details of Wallis' construction.

colleagues for his campaign to enter Parliament as member for Cambridge in 1705. However, due to his towering stature in English science, the text was soon translated from its original Latin, was widely read and became influential in Britain and on the Continent.

Despite Newton's evident discontent with details of the publication, the definition of number given in this work provides a concise synthesis of earlier conceptions as follows:

'By Number we understand, not so much a Multitude of Unities, as the abstracted ratio of any Quantity, to another Quantity of the same Kind, which we take for Unity. [Number] is threefold; integer, fracted, and surd, to which last Unity is incommensurable.' (Raphson's 1720 English translation, page 2.)[6]

This is the clearest statement yet of an approach that defines numbers as *abstract entities*. They are not taken as quantities, but may represent either quantities or ratios of the same. The Greek distinction between multiples and magnitudes is no longer an issue, and both rational and irrational numbers appear on the same footing. The same applies to negative numbers, where Newton does not follow Wallis and others in worrying about the philosophical implications of being 'less than Nothing', but draws analogies with debts and shortfalls, and works directly with positive and negative outcomes of a calculation in the same vein. Moreover, by treating 'surds' as numbers, Newton's classification moves us closer to the modern concept of 'real number'.

His attitude to square roots of negative numbers, on the other hand, seems ambiguous. He recognised that where such 'impossible' numbers appear as solutions of a polynomial equation, they should be accepted as genuine solutions, although he may have treated their occurrence in particular problems as having no clear real-world applicablity. In any event, they were not classed as *numbers* in the above definition, and their status was only resolved more than a century later.

In their different ways, decimal expansions, logarithms and analytic geometry involved ideas that were 'in the air' at the time of their invention – much the same is true of the Calculus. Those now credited with these achievements were usually the first to *publish* comprehensive results (although, as we saw in the cases of Jobst Bürgi and Pierre de Fermat respectively, others had simultaneously, or even earlier, developed similar concepts). The initial published results were developed and sometimes improved by their peers. Newton's famous comment (referring to the work of Descartes) in a letter to Hooke in 1676: *'If I have seen further it is by standing on the shoulders of giants'* (Newton wrote *'sholders'*) appears to be fully justified in this context.

---

[6]In the 1769 edition of Raphson's translation the following is added, presumably for emphasis: *'An Integer, is what is measured by Unity; a fraction, that which a submultiple Part of Unity measures; and a Surd, to which Unity is incommensurable.'*

3. Imaginary roots and complex numbers

But let us step back a little and return to Descartes. In the final book of *La géométrie* he turned to general principles for solving algebraic equations. He recognised that a polynomial of degree n,

$$p(x) = a_n x^n + a_{n-1} x^{n-1} + \ldots + a_1 x + a_0,$$

is divisible by $(x - \alpha)$ exactly when α is a *root* of the polynomial, that is, a solution of the equation $p(x) = 0$. He proceeded to argue that this implies that a polynomial of degree n has n roots (foreshadowing what we now call the Fundamental Theorem of Algebra): '*Every equation can have as many distinct roots (values of the unknown quantity) as the number of dimensions of the unknown quantity in the equation*'. Exactly what sort of 'numbers' (or 'values') should be allowed to represent these roots is not made explicit, although Descartes was surely aware that equations such as $x^2 + 1 = 0$ have *no* roots among the rational or irrational numbers represented by points of the geometric 'number line'.

3.1. The rule of signs. Descartes also expounded his *rule of signs*, which provides information on the number of positive roots of a polynomial (with multiply occurring roots counted by the number of times they appear—their *multiplicity*).

A variation in sign occurs in a polynomial

$$a_n x^n + a_{n-1} x^{n-1} + \ldots + a_1 x + a_0$$

if two consecutive coefficients have opposite signs. For example, $x^2 - 3x + 2$ has two variations: reading from the left we go from $+1$ to -3 to $+2$.

In modern terminology, *Descartes' rule of signs* can be stated as follows: the number of positive roots of a polynomial, each counted as often as its multiplicity, either *equals* the number of variations in the signs of its coefficients or is *less* than this number by an *even number*.

For example: the polynomials $x^3 - 4x^2 + 5x - 2$ and $x^3 - 3x^2 + x - 3$ each have three variations in sign ($+$ is followed by $-$, then $+$, then $-$). The rule of signs states that the positive roots of these polynomials will number either 3 or 1. Factorising each polynomial by inspection (trying out $x = 1, 2, 3$, for example) it is easy to see that the first can be written as $(x - 1)^2(x - 2)$, so the positive roots are $1, 1, 2$ (the repeated root is counted twice). The second, however, becomes $(x^2 + 1)(x - 3)$, so 3 is the only positive root, since $x^2 + 1$ has no real root.

Descartes himself formulated his rule of signs rather less clearly than stated here, for which Wallis took him to task in his *Algebra* (published in 1685). Wallis stated that Descartes had claimed that the number of positive roots would always *equal* the variation in signs, and remarked pointedly that the rule fails in general: '*it must be taken with this caution, that is, that the roots are real, and not imaginary'.* In Descartes' defence, we might observe that his claim was only that the number of positive roots *can* equal the variation in signs, and that his text appears to suggest that he was aware that there are fewer when imaginary roots occur.

3.2. Representation of imaginary numbers. These quotations suggest that, a century after *Bombelli's* struggle to make sense of them, imaginary roots of polynomial equations were no longer simply disregarded, but had become a possible object of study. In a further example of calculation, working implicitly with the 'positive square root' of -1, Leibniz factorised the expression $x^4 + a^4$ as

$$(x + a\sqrt{\sqrt{-1}})(x - a\sqrt{\sqrt{-1}})(x + a\sqrt{-\sqrt{-1}})(x - a\sqrt{-\sqrt{-1}}),$$

although he did not attempt to simplify the even more mysterious quantity $\sqrt{\sqrt{-1}}$ any further.

In another computation, Leibniz worked directly with the square root of -3 to obtain

$$\left(\sqrt{1 + \sqrt{-3}}\right)\left(\sqrt{1 - \sqrt{-3}}\right) = \sqrt{[1 + \sqrt{-3})][1 - \sqrt{-3}]} = \sqrt{1 - (-3)} = 2,$$

where, under the outer square root sign, he employed the familiar identity $x^2 - y^2 = (x + y)(x - y)$ with $x = 1, y = \sqrt{-3}$. Multiplying out the square $(\sqrt{x + y} + \sqrt{x - y})^2$ produces

$$\left(\sqrt{1 + \sqrt{-3}} + \sqrt{1 - \sqrt{-3}}\right)^2$$
$$= (1 + \sqrt{-3}) + (1 - \sqrt{-3}) + 2(\sqrt{1 + \sqrt{-3}})(\sqrt{1 - \sqrt{-3}})$$
$$= 2 + 2(2) = 6.$$

The identity $\sqrt{6} = \sqrt{1 + \sqrt{-3}} + \sqrt{1 - \sqrt{-3}}$ followed by taking square roots. This again showed that calculations with imaginary numbers could generate real numbers. Precisely what an imaginary number should signify remained obscure, although Leibniz and his successors encountered them ever more frequently in their studies of the Calculus and differential equations. Leibniz summed up the prevailing attitude to imaginary numbers in the 1680s as follows: *'From the irrationals are born the impossible or imaginary quantities whose nature is very strange but whose usefulness is not to be despised'.* [10].

Just over a century later, the concept of geometrical representation of imaginary numbers was very much in the air, and a subject of some controversy, as illustrated by the critical reception in England of a paper by the French émigré clergyman, *Adrien-Quentin Buée* (1745-1825), who maintained that $\sqrt{-1}$ should be seen as *'a purely geometric operation. It is a sign of perpendicularity'.*[7] In fact, within a period of less than fifteen years, geometric representations intended to represent *numbers* of the form $a + \sqrt{-1}b$, where a, b are real numbers, appeared independently in three different European countries. In 1797 the Norwegian surveyor *Caspar Wessel* (1745-1818) was exploring ways of representing *directed line segments*—segments of lines pointing in a given direction, defined by their length and their direction—that we call *vectors* today. Wessel was led by his considerations to represent $\sqrt{-1}$ as a *vertical* line segment of unit length, starting at the origin, in the Cartesian plane. He recognised that his geometric definition of the sum of two directed line segments (see Figure 23(a)) applied if the axes of his coordinate system are taken to represent real and imaginary numbers respectively. Wessel's paper, published in Danish, apparently remained unknown to most mathematicians of his day and only came to wider attention once it was translated into French a century later.

Another significant and more widely known advance was made by a Paris accountant and amateur mathematician by the name of Argand. Elementary textbooks still interchangably use the names *complex plane* or *Argand diagram* for the plane described by means of two perpendicular coordinates (as in Figure 23(b)), where the real numbers are on the horizontal axis and the imaginaries on the vertical one, containing i at one unit above the origin. However, very little is known reliably about the man himself—a somewhat dubious 1874 biography names him as *Jean-Robert Argand* (1768-1822) and places him as born in Geneva, but none of this has been verified by original sources.

Argand considered the imaginary unit in an essay published in 1806, regarding it as the result of a *rotation in the plane* through a right angle. He argued that 1 is transformed into -1 by rotating the plane through $180°$, and concluded that a rotation through half of this angle should lead to $\sqrt{-1}$ instead.

In the Argand diagram $a + ib$ is the point reached by moving a distance a from the origin along the real axis and then a distance b parallel to the imaginary axis. It was the third member of the trio (and its sole mathematician) Carl Friedrich Gauss, who, in notes and various publications dating from 1811 to 1831, pointed out that this represents the *complex number* $a + ib$ as a point (a, b) in the resulting plane. The resulting rectangular coordinate system has therefore also become known as the *Gaussian plane*.

Gauss remarked that, by dividing the plane into a grid by parallel lines, horizontal and vertical and one unit apart, the vertices of the resulting squares would become natural reference points for integral distances in both directions, with each point having four immediate neighbours.

---

[7]See the entry under his name in the MacTutor website (mathshistory.st-andrews.ac.uk), for example.

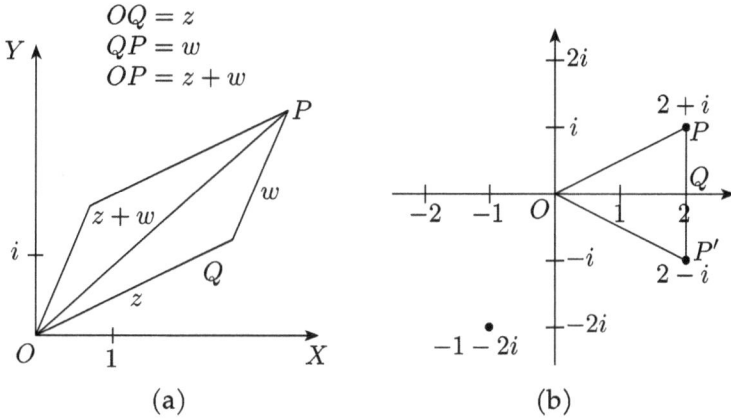

Figure 23. Representing complex numbers

Starting at the origin $(0,0)$, the four points on the axes a unit distance away are

$$(1,0),(0,1),(-1,0),(0,-1),$$

and he designated these as $1, i, -1, -i$ respectively. Gauss concluded:

'If this subject has hitherto been considered from the wrong viewpoint and thus was found to be enveloped in mysterious darkness, it is largely an unsuitable terminology which should be blamed. If $+1, -1, \sqrt{-1}$ had been described, verbally, not as 'positive, negative, imaginary' (or [the latter] even as 'impossible') but, for example, as 'direct, inverse, lateral' instead, there would have been no cause to refer to any such darkness'.

In a celebration commemorating the 50th anniversary of his Brunswick doctorate in 1849 Gauss was introduced with the phrase *'You have made the impossible possible'*[36].

3.3. Hamilton's definition of complex numbers. In the 1830s it was made explicit by the Irish mathematician *William Rowan Hamilton* (1805-1865) that a complex number could be described quite precisely as a *pair* (a, b) of real numbers, representing the Cartesian coordinates of a point in the plane. He argued that this should be its *definition* as a number. The geometric interpretation leads naturally to the definition of the *sum* of two complex numbers, which is found by adding coordinates separately. This allowed Hamilton to extend algebra to this set of numbers, although the *product* of two such numbers still required definition.

Hamilton approached *number systems* in a novel way: he sought to define complex numbers purely in terms of real numbers, without recourse to notions of 'imaginary' units or geometrical representations, but simply by setting down *arithmetical rules* for combining such numbers in a way that would be consistent with previously established properties.

For this, he assumed that the arithmetic of real numbers was unproblematic, and sought to recover known properties of complex numbers without trying to establish their 'true nature'. Instead, his arithmetical rules should *determine* the properties of the system.

As mentioned above, *addition* of two pairs of reals (a, b) and (c, d) could be defined simply as

$$(a, b) + (c, d) = (a + c, b + d),$$

since this would reflect the sum $(a + ib) + (c + id) = (a + c) + i(b + d)$ of two complex numbers. To define subtraction, one need only note that $(0, 0)$ acts as a neutral element for addition (i.e. $(a, b) + (0, 0) = (a, b)$), so that the additive inverse of (a, b) is $(-a, -b)$ since $(a, b) + (-a, -b) = (0, 0)$). More generally, the difference between two pairs is

$$(a, b) - (c, d) = (a - c, b - d).$$

To introduce *multiplication* Hamilton needed to reflect the property $i^2 = -1$ of the 'imaginary unit' i. This yields

$$(a + ib) \times (c + id) = (ac - bd) + i(ad + bc).$$

The *product* of two pairs of real numbers was therefore *defined* by Hamilton as

$$(a, b) \times (c, d) = (ac - bd, ad + bc).$$

To define division, he made use of the fact the quotient of two whole numbers, $\frac{m}{n}$, is the product of m with the multiplicative inverse $\frac{1}{n}$ of n. So the important step was to find the inverse, the number x satisfying $n \times x = 1$. The same idea can be applied to complex numbers, where the neutral element for multiplication is shown by the above definition to be $(1, 0)$. So the inverse (x, y) of (c, d) should satisfy

$$(1, 0) = (c, d) \times (x, y) = ((cx - dy), (cy + dx)).$$

Thus $cy + dx = 0$, hence $y = -\frac{d}{c}x$, and $1 = cx - dy = (c + \frac{d^2}{c})x$, so that the multiplicative inverse of $(c, d) \neq (0, 0)$ becomes

$$(x, y) = (\frac{c}{c^2 + d^2}, -\frac{d}{c^2 + d^2}).$$

Hamilton could therefore define the quotient of the pair (a, b) by the pair $(c, d) \neq (0, 0)$ as

$$(a, b) \div (c, d) = (a, b) \times (\frac{c}{c^2 + d^2}, \frac{-d}{c^2 + d^2}).$$

There is no distinction between the status of 'real' numbers $(a, 0)$ and 'imaginary' numbers $(0, b)$ in this system. Moreover, the multiplication

$$(0, 1) \times (0, 1) = (0 \times 0 - 1 \times 1, 1 \times 0 + 0 \times 1) = (-1, 0)$$

vindicates Bombelli's speculative 'wild thought' that led him to his multiplication tables for $\sqrt{-1}$.

Hamilton's definition of the *complex number system*, which we today denote by $\mathbb{C}$, states that it comprises the collection of all *pairs* (a, b) of real numbers with the two operations of addition and multiplication as defined above, noting that $(0, 0)$ and $(1, 0)$ would act as *neutral elements* for these operations respectively, that each pair (a, b) has an inverse under addition, and each pair $(a, b) \neq (0, 0)$ will have an inverse under multiplication, so that division of two pairs is defined as above.

Hamilton's approach does not require any pre-conceived notion of what complex numbers *are* or should 'represent'. His rules for combining complex numbers determine the *properties* of this number system, obviating any need to determine its 'underlying nature'.

In Figure 23(a) the addition of two complex numbers was depicted pictorially, while Figure 23(b) introduced the *conjugate* of $2 + i$ as its mirror image $2 - i$ in the horizontal axis. Both pictures are easily translated into Hamilton's definitions: $z = (a, b)$ has conjugate $\bar{z} = (a, -b)$, while its *modulus* $|z| = \sqrt{a^2 + b^2}$ is the radius of the circle about the origin containing z. It is immediate from the above definition that $z\bar{z} = (a, b) \times (a, -b) = (a^2 + b^2, 0) = |z|^2$, confirming that $|z|$ is the positive square root of $a^2 + b^2$. (These elementary facts will be used in the next section.)

Hamilton famously tried to extend his ideas to higher dimensions, i.e. triples and quadruples of reals, failing with the first and eventually succeeding with the latter in his invention of *quaternions*, which led to various unexpected applications. (See [6], p.220, for example.)

Hamilton's work initiated what became the modern approach to algebraic structures. This became a particular feature of nineteenth-century British mathematics (and symbolic logic), involving such figures as *George Boole* (1806-64), *George Peacock* (1791-1858) and *Augustus De Morgan* (1806-1871).

4. The fundamental theorem of algebra

Recall that Wallis had clarified Descartes' assertion about the number of roots of a polynomial of degree n: one can only expect n roots if complex numbers are allowed as roots. Neither of them proved this claim, and throughout the eighteenth century a variety of prominent mathematicians

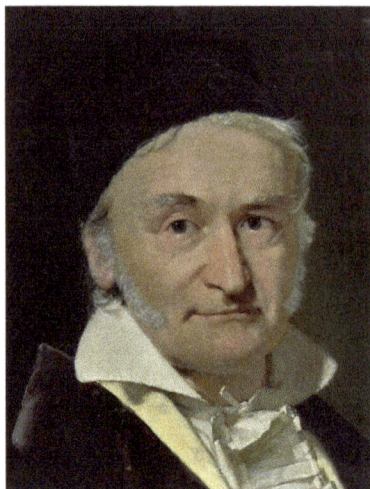

Figure 24. Carl Friedrich Gauss by C. A. Jensen, 1840[8]

set out to provide the proof. The first proof to gain wider acceptance appeared in 1799 in Gauss' doctoral thesis, which immediately established him as a major figure in his subject.

In his thesis, Gauss reviewed the attempted proofs of his predecessors, pointing out that all had *assumed* that any polynomial must have roots, although they did not specify which number system would contain these roots! This overlooked the main issue, he maintained, which was to demonstrate that any polynomial will have *at least one root* in some well-defined number system. His claim was that this is true if the number system in question is the complex plane. Thus the key result needed was the following, which we will call

Gauss' Theorem

Any polynomial $z^n + a_{n-1}z^{n-1} + ... + a_1z + a_0$ with complex coefficients a_i $(i = 0, 1, 2, ..n - 1)$ will have at least one complex root.

A polynomial with leading coefficient $a_n = 1$ is called *monic*. Restricting attention to this case is no real restriction. If the degree of a polynomial $c_nz^n + c_{n-1}z^{n-1} + ... + c_1z + c_0$ is n, the coefficient c_n must be nonzero, so that we can divide whole the polynomial by c_n to obtain a monic polynomial.

With Gauss' theorem we can can show that any monic polynomial of degree n is a product of n linear factors:

Fundamental Theorem of Algebra

---

[8]https://commons.wikimedia.org/wiki/File:Carl_Friedrich_Gauss.jpg

Any polynomial with unit leading coefficient,

$$p(z) = z^n + a_{n-1}z^{n-1} + ... + a_1z + a_0$$

and complex coefficients a_i ($i = 0, 1, 2, ..n - 1$) has n (not necessarily distinct) complex roots $\alpha_1, \alpha_2, ..., \alpha_n$, and can therefore be factorised (uniquely, up to the order of factors) into n linear factors, so that

$$p(z) = (z - \alpha_1)(z - \alpha_2)...(z - \alpha_n).$$

Proof

By Gauss' theorem there is a complex number w such that $p(w) = 0$. Assuming this result we can now write

$$p(z) = p(z) - p(w) = (z^n - w^n) + a_{n-1}(z^{n-1} - w^{n-1}) + ... + a_1(z-w) + a_0(1-1).$$

In particular, the constant term $a_0 - a_0 = 0$, and for any $k = 1, 2, .., n - 1$ we can factorise (just as we did for $z^n - 1$ earlier)

$$z^k - w^k = (z - w)(z^{k-1} + z^{k-2}w + ... + zw^{k-2} + w^{k-1}).$$

This shows that $(z - w)$ is a common factor of all the remaining terms in the expansion of $p(z) = p(z) - p(w)$, so that we can re-arrange terms to write $p(z) = (z - w)q(z)$ for some polynomial q of degree $(n - 1)$. Write $\alpha_1 = w$, and use Gauss' theorem to find a root w' of the polynomial q. As before, $(z - w')$ is then a common factor of $q(z) = q(z) - q(w')$, so we can find a polynomial r of degree $(n - 2)$ such that $q(z) = (z - w')r(z)$. Write $\alpha_2 = w'$. We have shown that $p(z) = (z - \alpha_1)(z - \alpha_2)r(z)$.

Continuing the same process we can find roots $\alpha_3, ..., \alpha_n$ of p, so that, finally, $p(z) = (z - \alpha_1)(z - \alpha_2)...(z - \alpha_n)$.

This completes the proof of the theorem.

This illustrates how what we have called Gauss' theorem is the key result required for understanding the structure of polynomials. A different proof of the Fundamental Theorem was published by Argand in 1814. His proof was not deemed rigorous and it was not widely accepted at the time – although it now thought that Argand's ideas provide the most direct approach to the problem. In fact, all the attempted proofs published in the early 1800s could only be made fully rigorous in the 1870s, when the crucial role played in these arguments of *completeness* of the real number system – which we discuss in **Chapters 7** and **8** – was better understood.

Argand's ideas are based on a result first announced by the French mathematician *Jean le Rond d'Alembert* (1717-1783), who had also made two attempts (1746 and 1754) to prove the Fundamental Theorem.

d'Alembert's lemma:

If the complex polynomial $p(z)$ is non-constant and $p(z_0) \neq 0$, then any neighbourhood of z_0 contains a point w with $|p(w)| < |p(z_0)|$.

In his papers d'Alembert takes for granted that, as z varies across the complex numbers, the curve traced out by $|p(z)|$ will be *continuous*. In his time, continuity was described in terms of 'infinitesimal change', so that a continuous curve was regarded as one that could be drawn without lifting the pencil, and the notion of 'neighbourhood' relied on geometric intuition. In the nineteenth century it became clear that these ideas do not provide adequate definitions.

There are many modern proofs of versions of Gauss' theorem. Those making direct use of d'Alembert's result are perhaps the most accessible. A relatively brief, but authoritative, summary of the history of this important result and its many different proofs is given in Chapter 4 of [10]. A short modern proof is given in *MM*.

Today the Fundamental Theorem of Algebra can arguably be regarded as mis-named, since it deals exclusively with polynomial equations, which are far removed from the myriad abstract algebraic structures that have been invented in the two centuries since Gauss' heyday. Moreover, its early proofs implicitly assumed deeper properties of the number line than could be made visible by eighteenth century algebra. The fact that Gauss returned to the theorem three times throughout his career testifies that he was aware that his original argument in 1799 contained a significant gap. It was only in 1920 that the Russian mathematician *Ostrowski* fully completed Gauss' original proof, in line with modern standards of rigour.

Gauss' theorem is a good example of the perspectives of modern mathematics which he did much to encourage. The focus is not on finding a construction of the actual roots, but on showing that, in general, roots of polynomials will always exist. It is proved that they are there to be found, without actually specifying how to find them. By shifting perspectives in this way, mathematicians found that much larger and previously intractable, even unimagined, areas of enquiry became available. The focus now shifted to analysing the *structure* of the objects (here, the collection of all polynomials) being investigated. From this point of view Gauss' theorem is fundamental, since it shows that, as long as we are flexible about the kinds of *numbers* we allow as roots, then the structure of any polynomial is fully understood once we know all its roots.

CHAPTER 5

Struggles with the Infinite

Again there is another great and powerful cause why the sciences have made but little progress; which is this. It is not possible to run a course aright when the goal itself has not been rightly placed.

Sir Francis Bacon, *Novum Organum*, 1620

Summary

In this chapter we review how mathematicians (and sometimes philosophers) of previous centuries dealt with the troublesome concept of *infinity*. Our overview must necessarily be concise – this is not a full historical account by any means! We will focus, instead, on two key periods, nearly 2000 years apart.

For the first we return to Ancient Greece to consider Aristotle's conception of the *potential infinite* and the difficulties that notions of infinite divisibility of space and time presented. Next come the works of Archimedes, also transmitted via the Arab world, with their remarkably sophisticated comparisons of the areas and volumes of various curvilinear figures. One might echo Descartes' suspicions (mentioned in **Chapter 3**) *'that these writers then with a sort of low cunning, deplorable indeed, suppressed'* their methods for *discovering* these relationships. In fact, Archimedes' recently rediscovered letter, *The Method of Mechanical Theorems*, addressed to his friend *Erastosthenes*, shows how he had used his *law of the lever*, together with 'infinitesimal slices' of solid bodies and areas, to arrive at his results. These ingenious techniques, foreshadowing arguments used nearly two millennia later in the Calculus, did not conform to the rigorous standards of proof of Euclidean geometry. In his public tracts Archimedes stated his results and proved them *by contradiction*, often generalising the 'method of exhaustion' established earlier by Eudoxus, with no explanation how he had discovered the relationships his proofs verified.

Finally we consider the development of the Calculus from the late seventeenth century onward, including the serious logical issues it posed. We focus on the different conceptions of the two principal contributors, Isaac Newton and Gottfried Wilhelm Leibniz. While they are rightly celebrated

https://doi.org/10.11647/OBP.0236.05

as inventors of the Calculus, both relied on versions of a controversial 'principle of continuity', expressed by Leibniz as *'whatever succeeds for the finite, also succeeds for the infinite'*. The logical difficulties this created are seen most directly in the methods of proof employed in the eighteenth century in the prolific writings of Leonhard Euler – his results were almost always correct, but many could only be verified rigorously in the nineteenth century. His treatment of the 'natural logarithm' function is a classic example.

1. Zeno and Aristotle

Differing perceptions of the nature and role of infinity in mathematics have pervaded the subject since the era of Ancient Greece. Aristotle insisted that mathematicians have no need of *actual infinity*, but can work with the unlimited, the *potentially infinite*. For example, the natural numbers may be regarded as a potentially infinite collection, since the process of counting them can in theory be continued indefinitely, even though, in practice, we cannot continue counting forever (as I found out and reported in the **Prologue**). His perceptions, although highly influential, have not always held sway.

Early in the fifth century BCE, the philosopher Parmenides of Elea reasoned that nothing can ever change, because *'nothing comes from nothing'*. In support of this view, *Zeno* (490-430 BCE) presented a number of famous *paradoxes*. He wished to show that motion is logically impossible, and is therefore a sensory illusion. To do this, he presented arguments that challenge the notion of the 'infinite divisibility' of space and time. His paradoxes stimulated much discussion among the Pre-Socratic Greek philosophers, and were addressed at some length by Aristotle and other later commentators.

The most famous of his examples is popularly known as *Achilles and the Tortoise,* although Zeno merely asserts that in a race the quickest runner cannot overtake the slowest if the latter has a head start.[1] In the popular version, Achilles, the mythical fastest runner in antiquity, cannot overtake a slow tortoise: to reach the tortoise he must first pass the tortoise's starting point A, by which time the tortoise has moved to some point B, further on. When Achilles reaches B the tortoise is at some point C beyond B, and so on indefinitely. Thus, while the tortoise's lead becomes ever smaller, after any finite number of these stages it is still ahead. To overtake the tortoise, Achilles would have to cover an infinite number of intervals in a finite time, which is impossible, Zeno argued.

Aristotle struggled to refute Zeno's reasoning conclusively. He argued that Achilles moves in a 'continuous' motion, thereby implicitly conceding

---

[1]Aristotle renders the claim as follows in his *Physics:* In a race, the quickest runner can never overtake the slowest, since the pursuer must first reach the point whence the pursued started, so that the slower must always hold a lead.

that time and space can be regarded as potentially infinitely divisible. He says that an infinite number of intervals can be covered in a finite time, since *'while a thing in a finite time cannot come in contact with things quantitatively infinite, it can come in contact with things infinite in respect to divisibility, for in this sense time itself is also infinite'*. This does not explain how we might compute the instant at which Achilles will overtake the tortoise. Today, this can be done with an apparently (!) simple calculation.

To fix ideas, assume that the tortoise has a head start of 10 units and moves at 1 unit per minute, while Achilles is 10 times as quick, so that he covers each unit in $\frac{1}{10}$th minute. Achilles reaches the tortoise's starting point in 1 minute, by which time the tortoise has moved on 1 unit. Achilles covers that unit in the next $\frac{1}{10}$th minute, at the end of which the tortoise is now only $\frac{1}{10}$th unit ahead, which Achilles then covers in $\frac{1}{10} \times \frac{1}{10} = \frac{1}{10^2}$ minutes, and so on.

Hence the time (in minutes) it takes Achilles to catch up with the tortoise is given by the sum of the infinite geometric series

$$1 + \frac{1}{10} + \frac{1}{10^2} + \dots + \frac{1}{10^n} + \dots$$

However, this is a series with infinitely many terms, so we need to understand what we mean by its sum. For any finite n we can sum the first n terms, and consider what happens when we let n 'grow'. But we need to decide what this last phrase should mean.

More generally, let $-1 < x < 1$ and set $S_n = 1 + x + x^2 + \dots + x^{n-1}$. We call S_n the nth *partial sum* of the series $1 + x + x^2 + \dots + x^n + \dots$ If we now multiply both sides by $(1 - x)$, we find that

$$(1 - x)S_n = (1 - x)(1 + x + x^2 + \dots x^{n-1}) = 1 - x^n,$$

as the inner terms in the product cancel in pairs. This is the archetypal example of a 'telescoping sum'. So we have

$$S_n = \frac{1 - x^n}{1 - x} = \frac{1}{1 - x} - \frac{x^n}{1 - x}.$$

For fixed $x < 1$, $\frac{1}{1-x} > 0$ is constant, and the final term will behave like x^n. We claim that x^n can be made as close to 0 as we please by taking n large enough – a formal proof of this will be given in **Chapter 7**. Assuming this for now, we see that the sequence of these partial sums $(S_n)_{n \geq 1}$ therefore gets ever closer to $\frac{1}{1-x}$. This 'limiting value' is now taken as the *sum* of the infinite series.

When $x = \frac{1}{10}$, we obtain $\frac{1}{1 - \frac{1}{10}} = \frac{10}{9}$, so that Achilles catches up with the tortoise after $\frac{10}{9}$ minutes.

This kind of quantitative analysis was not available to Aristotle. He rejected the Pythagorean claim that 'All is Number' and argued that there are two types of quantities, distinguishing between discrete *multiples* (of the unit) which could be represented by whole numbers (where each number has an immediate successor), and continuous *magnitudes* (where there are *no* immediate successors). Multitudes could be handled with arithmetic, while magnitudes belonged to the domain of geometry. For Aristotle, objects like line segments with a common endpoint touch each other and, as he puts it, *'the touching limits of each become one and the same'.* He argues that a continuous line, a *continuum*, should not be seen as an aggregate of individual 'points'. Successive division of a line segment into two equal parts produces not points, but ever shorter line segments, which eventually become smaller than any pre-assigned magnitude.

Aristotle insists that continuous motion 'flows' along the line and cannot be described as going from point to point in succession. Thus points only have potential, not actual, existence and adding them together does not produce a line segment. In his *Physics* he argues that division of a continuous line into two halves makes the original midpoint into an endpoint in each half, which destroys continuity, both of the motion and of the line. The line can be halved repeatedly to produce an unbounded number of such successively shorter halves, not in reality but only potentially. It is a process whose end result is never fully actualised. For Aristotle, *continuity* of motion (as yet undefined mathematically) is the key concept.

The emphasis on describing these ideas by means a static analysis of shapes, as in plane and solid geometry, may explain why the Ancient Greek mathematicians had difficulty in describing motion mathematically. Aristotle insisted that a given velocity achieved by a moving body must *'persist for a time'*, and rejected notions of 'instantaneous' change. This, together with the distinction between the discrete and continuous, meant that Greek mathematicians were not able to deal effectively with variable motion.

The key stimulus that would provide a solution to these problems was the gradual development of the Calculus in the sixteenth and seventeenth centuries – although, as will be seen next, a good deal of this was foreshadowed in a work of Archimedes which remained lost for nearly a millennium and was not analysed fully until quite recently. Between 1200 and 1600, thinkers in various parts of Europe extensively debated Aristotle's views when discussing variable motion as well as the nature of space and time. The notion of the infinite divisibility of space and time gradually gained ground. Unknown to them, many of their ideas had been foreshadowed by Archimedes.

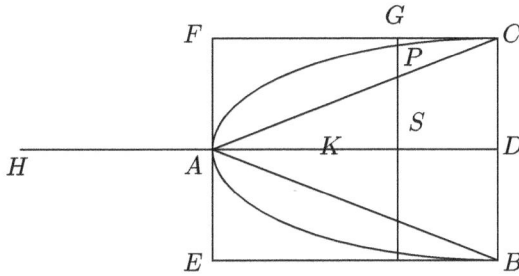

Figure 25. Paraboloid of revolution and cone

2. Archimedes' 'Method'

A century after Aristotle, Archimedes developed tools for the calculation of areas and volumes of various curvilinear bodies (sphere, cone, cylinder, parabolic segment, spirals, etc.), leading to startling new numerical relationships between these objects.

Two typical examples, one comparing volumes and one comparing areas, illustrate the remarkable sophistication and scope of his results:

The first is from his treatise *On Conoids and Spheroids*:

Theorem

The volume of a segment of a paraboloid of revolution cut off by a plane at right angles to the axis (we might think of this as a 'bullet') *is in the ratio 3 : 2 to that of the cone* (think ice-cream cone!) *which has the same base and axis.* (See Figure 25.)

In his surviving works, probably housed first in the library of Alexandria, later translated or transcribed in Baghdad and Constantinople before reaching the West some 1500 years after his death, Archimedes stated many results like the above, and invariably verified the formulae he discovered by extending the *method of exhaustion,* due to Eudoxus more than a century earlier and used extensively by Euclid. This method involves using inscribed and circumscribed figures, most often regular polygons or circular segments, whose properties were well understood, fitting inside and outside the given shape. The inscribed and circumscribed figures are then successively modified (typically by bisecting the sides of regular polygons) in order to approximate the desired shape ever more closely. One then confirms the truth of the given formula with a proof by contradiction—assume that one side of the equation to be verified is greater than the other, then show that, in finitely many steps, we will arrive at a claim that contradicts a known property of the approximating figures.

In works such as *On Conoids and Spheroids* Archimedes gives no indi-cation how he arived at his results in the first place—after all, a proof by contradiction requires him to assume that the ratio to be verified (3 : 2 in this case) is *incorrect* and then to show that this leads to a contradiction.

However, in a letter entitled *The Method of Mechanical Theorems* (now usu-ally simply called *The Method*) and addressed to Eratosthenes, Archimedes described clearly how he used his well-known *law of the lever* to compare *in-finitesimal slices* of solid bodies to arrive at the formulae he then proceeded to verify painstakingly, using (and extending) Eudoxus' method of exhaus-tion.[2] The argument in the shaded paragraph below shows how he relates the volume of the circular cone to that of the cut-off paraboloid with the same base, as stated above.

In Figure 25, consider the segment BAC of a parabola with vertex at A and cut off by the line BC. We compare segment BAC with the rectangle $CBEF$ and with the triangle ABC. Draw AD parallel to FC and EB, so that, by the symmetry of the parabola, D is the midpoint of BC. Rotating all three plane shapes—the trian-gle ABC, the parabolic segment and the rectangle $CBEF$—through a full revolu-tion about the line AD produces a circular cone, a paraboloid of revolution (which we called a 'bullet' in the statement of the theorem) and a cylinder, respectively. The cone will have volume equal to $\frac{1}{3}$ that of the cylinder, as was well-known in Archimedes' time.

Now extend DA to H so that $DA = AH$, and, from now on, treat A as the fulcrum of a lever. Archimedes imagines himself 'weighing' infinitesimal slices of the 'bullet'—represented by lines in our two-dimensional figures–against slices of the cylinder, placed where they are on DA. Choosing any point P on the parabolic segment, he draws the line $GPSM$ parallel to CB to meet CF in G, the parabolic segment in P, DA in S and BE in M.

By its definition[a] the parabola satisfies the proportion $DA : AS = (CD)^2 : (PS)^2$. The line segments PS, SM, rotated about S on DA, produce circles with radii PS and $GS = CD$ respectively. As circles are to one another as the square on their radii, he concludes that $DA : AS$ is the same ratio as the area of the circle in the cylinder is to that of the circle in the paraboloid of revolution. But $HA = DA$ so the ratio $HA : AS$ has the same property.

So the circle of radius GS in the cylinder (thought of as an *infinitesimal slice*), placed where it is, will balance the circle of radius PS if the latter is placed at H. But we may regard the weight of the whole cylinder as being placed at its centre of gravity, which is the midpoint K of AD. Archimedes imagines placing the centres of all the circular slices of the cylinder at K, and 'balances' them against the totality of slices of the parabolic segment, all placed at H.

---

[2]The law of the lever states that on a scale with a central fulcrum and two linear arms (like a seesaw) the scale will balance precisely when the *product of* the weight placed on one side and its distance from the fulcrum equals the same product on the other side.

In *Conoids and Spheroids* he had shown that equality of ratios for all the individual pieces implies equality of the sums of n such ratios taken on each side, for every finite n. Now he *asserts* (without proof) that this also holds for sums of *infinitely many* ratios, and applies this to the sums of his 'slices'.

Consequently, by the law of the lever, the ratio $HA : AK$ represents the ratio of the cylinder and the segment of the paraboloid of revolution! Since $AK = \frac{1}{2}HA$, the volume of the parabolic segment is half that of the cylinder, and, since the cylinder has volume three times that of the cone, the volume of the segment is $\frac{3}{2}$ times that of the cone, as he had claimed.

[a]Today we would write this as $y^2 = 4ax$, meaning in particular that the change in the y-direction is proportional to the square of the change in the x-direction.

Archimedes' use of his 'law of the lever' shows that he treated the volume of the curved bodies he compares as proportional to their weight. He also assumes that the total volume of his infinitely many 'infinitely thin' slices will equal the volume of the whole body. These techniques, while entirely plausible, do not conform to the rigorous demands of Euclidean geometry. This explains why Archimedes then goes on to prove by the 'method of exhaustion' that the ratios he has identified using this technique are the correct ones.

On the other hand, as a means for *discovering* what these ratios must be, his informal arguments are clearly very productive. His use of summing infinitely many infinitesimal slices was well ahead of its time. It was eventually reinvented independently some 1800 years later to produce the integral calculus for the computation of areas and volumes.

As a second example of the fruits of Archimedes' novel techniques we briefly mention a result from his later work *On Spirals:*

The area of the first full turn of the spiral is $\frac{1}{3}$ of the area of the circle whose radius is the distance between the origin O of the spiral and the point P reached at the first full turn.

(In Figure 26, the shaded area is $\frac{1}{3}$ of the area of the circle with centre O and radius OP.)

The Archimedean spiral is defined as the locus of a point, starting from O, which moves uniformly along a line OA, which is itself rotating uniformly about O. Archimedes puts it as follows:

'If a straight line, one extremity of which remains fixed, be made to revolve at a uniform rate in the plane until it returns to its starting position and, if at the same time as the straight line is revolving, a point moves at a uniform rate along the straight line, starting from the fixed extremity, the point will describe a spiral in the plane.'

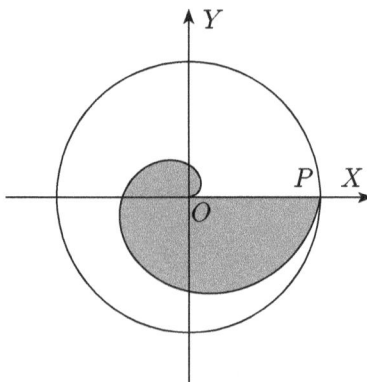

Figure 26. Spiral and circle

We express this using polar coordinates (see **Chapter 4**): $r = a\phi$ for some $a > 0$, where ϕ is the angle at O that OA makes with its original position. At P in Figure 26 we have $\phi = 360°$. The constant a is the ratio of the constant velocity of the linear motion of the point and the constant angular velocity of rotation of the line. Archimedes' quite involved constructions will be omitted here (see [19]).

Archimedes' use of the concept of the locus of a point involves *motion*, but the velocity of the moving points and lines used explain how to trace out complex figures (which typically cannot be drawn by rules and compass alone), is always assumed to be constant. Mathematical descriptions of the motion of *accelerated* objects, one of the key features of applied Calculus, had to wait for Isaac Newton.

One possible reason why it took so long may be that for fully a thousand years, between the tenth and twentieth centuries, Archimedes' *Method* disappeared from view, so that its contents were not available to Renaissance and seventeenth-century mathematicians, who had to reconstruct them from scratch, with much painful effort and over an extended period.

The Method seems to have had no influence on Islamic geometry, although many Greek manuscripts were studied minutely by scholars in Baghdad and Constantinople from the eighth century to the eleventh century. It could have been available to them, however. In the sixth century, *Isidorus of Miletus*, the architect of the *Hagia Sophia* church in Constantinople, collected, into a single document, letters by Archimedes previously held in the Great Library of Alexandria. This included what we now know as *The Method*. Around the middle of the tenth century an unknown scribe copied it onto a parchment that then found its way to Jerusalem by the thirteenth century, where the text was partially erased and overwritten by monks with Christian liturgical text and lost to science. There is no evidence that any copies of

The Method were transmitted to Europe during the Renaissance. The heavily overwritten palimpsest was discovered in a monastery and brought back to Constantinople around 1840, where it was catalogued.[3] Around the turn of the century it was studied *in situ* by the eminent Danish historian *Johan Heiberg*, who confirmed Archimedes' authorship. Amid the tumult of the Greco-Turkish war that followed World War I, the palimpsest then disappeared once more. It finally resurfaced at an auction in 1998, was bought for $2.2 million, and is now held in Walter's Art Museum, Baltimore, USA.[4]

3. Infinitesimals in the calculus

Mediaeval thinkers, meanwhile, had spent a good deal of effort on the problem of infinite divisibility of lines. For example, *Thomas Bradwardine* (1295-1349), Archbishop of Canterbury, argued that the line is made up of infinitely many *indivisible* segments, but distinguished these 'atoms' from 'points'. He argued that the atoms were magnitudes of the same kind as the line they produced – in line with Aristotle's thinking. This viewpoint allows 'infinitesimals' only a potential existence: the continued division of the line in arbitrarily many steps only produces ever shorter lines, not points.

Such arguments did little to resolve some of the obvious *paradoxes of the infinite* that gained more attention in Europe during the Middle Ages. For example, two concentric circles share the same lines for their radii, so the 'atoms' of their respective circumferences can be 'paired off' exactly, each pair consisting of the two points where a given radius meets the two circumferences. Yet the circumferences of the two circles are clearly not the same!

A paradox along similar lines is contained in the famous *Dialogue concerning two New Sciences* (1638) by *Galileo Galilei* (1564-1642): among (positive) whole numbers, some are perfect squares (the 'square numbers' of **Chapter 1**) and some are not. Therefore there should be more positive whole numbers than square numbers. Yet, for each $n \geq 1$, the positive whole number n can be paired uniquely with the nth perfect square n^2. So the collections of positive whole numbers and the perfect squares can be 'paired off' exactly. Despite evidence that this example had been known for some time

---

[3]A *palimpsest* is a document where earlier writing has been partially erased and overwritten.

[4]The later history of the Archimedes palimpsest is bizarre. During the 1920s it was acquired by a French traveller in the Middle East, having fairly recently been overpainted with gold leaf by a forger. It then spent more than 60 years (many of them in a mouldy cellar) with his family, who, initially unaware of its significance, had sought a private buyer for several years before putting it up for auction. After 1998, the text was studied extensively, translated and finally published by Cambridge University Press in 2011 as *The Archimedes Palimpest*, (2 vols.). Reviel Netz, one of the authors, claims (perhaps with a degree of hyperbole) that the palimpsest reveals that the work of the Western scientific revolution since the seventeenth century is, in essence, simply 'a series of footnotes to Archimedes'.

before Galileo, his fame ensured that this observation became widely known as *Galileo's Paradox*. It was variously discussed until the late nineteenth century.

Examples like these led to an early realisation that there are difficulties in handling *infinite aggregates* arithmetically. Nevertheless, notions about the usefulness of the infinitely small and infinitely large in mathematics increasingly found adherents, especially in connection with theological speculations and a renewed interest in Plato's philosophy, which theologians such as *St. Augustine of Hippo* (354-430) had earlier sought to reconcile with biblical dogma.

3.1. The Principle of Continuity. *Nicholas of Cusa* (1401-1461), born as Nikolaus Krebs in what is now Berncastel-Kues on the Moselle, became arguably the most influential German theologian and philosopher of the fifteenth century, serving as papal legate to Germany for much of his later career. He illustrated his notion of the *'coincidence of opposites'*, which he sees in the relationship between between God and Man, with various mathematical metaphors, arguing that:

(i) by continuously increasing the number of sides of a polygon, we will eventually reach a circle,

(ii) by increasing the radius of a circle indefinitely, the tangent at a point becomes identical with the circumference,

(iii) although the centre and circumference of a circle are opposites, by shrinking the radius until it is infinitesimal, these opposites coincide.

In this way he sought to reconcile the apparent contradiction of the finiteness of our world and the infinite being of God: the diversity and multiplicity of our finite existence become one in the realm of God, who both transcends and resides in every part of the universe.

His cosmology, based on these precepts, was remarkably prescient, although not based on any direct evidence. He believed that the universe has no 'centre' (rejecting prevalent geocentric doctrines well before Copernicus): for him, the universe and its centre are the same.[5] Nor is the Earth at rest: *'It is impossible for the world machine to have this sensible earth, air, fire, or anything else for a fixed and immovable centre. For in motion there is simply no minimum, such as a fixed centre.... And although the world is not Infinite, it cannot be conceived of as finite, since it lacks boundaries within which it is enclosed. ... Therefore,*

---

[5]In *De Docta ignorantia* he writes: *Life, as it exists on Earth in the form of men, animals and plants, is to be found, let us suppose in a high form, in the solar and stellar regions. Rather than think that so many stars and parts of the heavens are uninhabited and that this earth of ours alone is peopled – and that with beings perhaps of an inferior type – we will suppose that in every region there are inhabitants, differing in nature by rank and all owing their origin to God, who is the centre and circumference of all stellar regions*

just as the earth is not the centre of the world, so the sphere of fixed stars is not its circumference'.

Nicholas' thinking influenced the work of major mathematical figures such as Kepler, Leibniz and Euler, over the next three centuries. *Johannes Kepler* (1571-1630), for example, imagined the sphere of radius r as made up of an infinite number of infinitely thin cones with their vertices at the centre of the sphere, and bases on the surface of the sphere. He took their bases to be small enough to allow him to assume that the flat base of the cone and the surface area of the correponding part of the sphere are the same. He then calculated the volume of the sphere to be $\frac{4}{3}\pi r^3$. For this, he applied two facts well-known since Ancient Greece: the volume of each cone is $\frac{1}{3}$ of the product of its base and its height r, while the sum of the (infinitely many!) bases is the surface area of the sphere, which the Greeks had shown to be $4\pi r^2$.

Similarly, the Italian mathematician Bonaventura Cavalieri calculated the areas of various curved bodies by imagining their areas as made up of an infinite number of line segments (regarded alternatively, as 'infinitely thin' slices, to ensure that each had the same dimension as the total figure, as Aristotle had demanded) and summing this infinite collection to find the desired area.

3.2. Leibniz. In the 1670s, *Gottfried Wilhelm Leibniz* (1646-1716) formulated his version of the ideas initiated by Nicholas of Cusa as a formal *Principle of Continuity*. One may regard it essentially as an operational maxim: 'whatever succeeds for the finite, also succeeds for the infinite' – although, in terms of his mathematical ideas, he may well be referring to the infinitely *small* rather than the infinitely large! Expressing his principle more formally, he writes in 1701:

'In any supposed continuous transition, ending in any terminus, it is permissible to institute a general reasoning, in which the final terminus may also be included'.

Leibniz used this principle to justify his extensive use of *infinitesimal* quantities, which he employed to compute curvilinear areas (see Figure 28). This enabled him to turn Cavalieri's 'method of indivisibles' into a technique for finding the areas under various types of ('smooth' enough) curves.

Crucially, he realised early on in his studies that finding the *area* under a curve (*integration*) and determining the *tangent* to the curve (*differentiation*) were opposites, or *inverse operations*. A series of manuscripts in which he noted down his developing insights in late 1675 suggests how this came about.[6]

---

[6]This very brief summary draws on the essay *Newton, Leibniz and the Leibnizian Tradition* by Henk Bos, in [17].

Figure 27. Gottfried Wilhelm Leibniz by C.B. Francke, (d. 1729).[7]

The origin of his discovery lay in the simple relationship between a sequence of numbers $(a_i)_{i\geq 1}$ and the sequence $(b_j)_{j\geq 1}$ of their successive differences, $b_j = a_j - a_{j+1}$. In Paris three years earlier, while being guided into mathematical study by the Dutch mathematician, physicist and astronomer *Christiaan Huygens* (1629-1695), Leibniz had already noted that a difference sequence is easily summed, since it becomes what is now called a telescoping sum. Since for each j adding pairs of successive terms produces cancellations, i.e.

$$b_{j-1} + b_j = (a_{j-1} - a_j) + (a_j - a_{j+1}) = a_{j-1} - a_{j+1}$$

it follows that for any $n \geq 1$ we have $b_1 + b_2 + \dots + b_n = a_1 - a_{n+1}$.

He had applied this when Huygens asked him to sum the reciprocals of triangular numbers $1, 3, 6, 10, 15, 21, \dots$ Since a triangular number takes the form $\frac{1}{2}k(k+1)$, its reciprocal is $\frac{2}{k(k+1)} = \frac{2}{k} - \frac{2}{k+1}$. Taking $a_k = \frac{1}{k}$, he obtained $b_k = \frac{2}{k(k+1)}$. The partial sum $b_1 + b_2 + \dots + b_n$ of the first n terms of the series

$$\frac{1}{1} + \frac{1}{3} + \frac{1}{6} + \frac{1}{10} + \frac{1}{15} + \dots$$

became $a_1 - a_{n+1} = 2 - \frac{2}{n+1}$. As $\frac{2}{n+1}$ becomes infinitesimal for infinite n, Leibniz concluded that the sum of $b_1 + b_2 + \dots + b_n + \dots$ was 2.

Leibniz studied this and other examples in detail, coming to the realisation that the operations of forming difference sequences and sum sequences are in effect mutual inverses – each undoes the other. He applied this insight to geometric curves in the plane. Leibniz perceived a curve in the (x, y)-plane as depicting the values taken by a *variable quantity* y whose changes in value depend on changes in the value of the variable x. He considered the slope of the *tangent* to the curve (a straight line touching, but not crossing,

---

[7]https://commons.wikimedia.org/wiki/File:Christoph_Bernhard_Francke.jpg

$$dx_i = 1 \qquad\qquad \text{Area} = y_1 + y_2 + \dots + y_n$$
$$dy_i = y_{i+1} - y_i$$

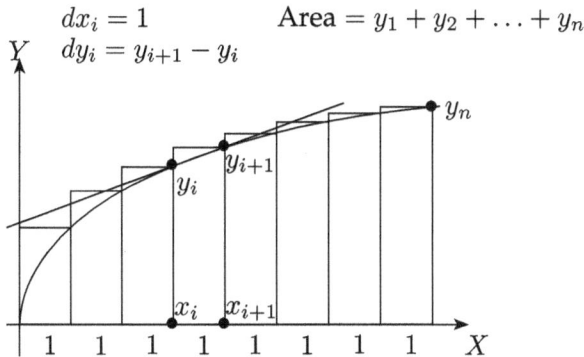

Figure 28. Leibniz' calculus

the curve) at various points in the (x, y)-plane. He treated this slope as the *ratio* of *infinitesimal increments* dy and dx in y and x respectively. In Figure 28, he then takes each increment in x as the *infinitesimal* 'unit', $dx_i = x_{i+1} - x_i$ (denoted here by 1), so that the tangent at the point (x_i, y_i) becomes the *difference* $dy_i = (y_{i+1} - y_i)$ of two successive ordinates.

The area under the curve, on the other hand, is taken as the *sum* of the areas of infinitely thin rectangles, with base vertices x_i and x_{i+1} on the x-axis and height given by the ordinate y_i. Since the base of each rectangle is infinitesimal, he assumes that the heights along the top edge of each small rectangle remain infinitely close to the corresponding part of the graph of the curve. The base of each rectangle is one infinitesimal unit, so the area under the curve becomes the *sum* of the ordinates: $y_i + y_2 + \dots + y_n$, where n is infinite.

Difference sequences and sum sequences are opposites. For Leibniz this illustrates the inverse relationship, for a given curve (denoted here by a function f), between *differentiation*, in which we find the values of its tangent curve (denoted by f') at various points, and *integration*, where we seek to express f in terms of the area under the graph of its tangent curve f'. The inverse relationship between these two basic operations became known as the *Fundamental Theorem of the Calculus*.

Leibniz' results represented a major step forward in the use of the new-found algebraic symbolism to describe properties of curves and move beyond the confines of Greek geometry in the study of accelerated motion. However, his methodology led to serious foundational questions about the *existence* of the objects being studied, since it was by no means clear how the infinitesimal quantities could serve as fundamental building blocks on which a rigorous logical foundation of the Calculus, in the pattern of Euclid's *Elements*, could ultimately be based.

Figure 29. Sir Isaac Newton by J. Faber junior, 1726.[8]

He was fully aware that the use of infinitesimal quantities needed to be justified. He had no proof of the existence of such entities, but regarded them as 'ideal quantities', to be used in a formal framework for calculation. Using his *Principle of Continuity*, he asserted that these fictional 'ideal' numbers were governed by the same laws as 'ordinary' numbers, by which he meant rational or irrational numbers, the existence of the latter being justified by an appeal to the geometric number line.

At the same time, he and his followers made the infinitesimal increment dx (called the *differential*) the basis of their calculations. They claimed that two quantities could be treated as the same if their difference was *infinitesimal* (smaller than any given positive quantity). The *instantaneous rate of change* in y at the point (x, y), which yields the slope of the tangent at the point, was then assumed to be given by the *ratio* $\frac{dy}{dx}$ of two infinitesimals.

The logical inconsistency of these claims was obvious – infinitesimals could not simultaneously by treated as zero and as non-zero quantities. Nevertheless, judicious and selective use of techniques for calculation based on the above premises provided an *operational* foundation for a highly successful Calculus with wide-ranging applications, which produced convincing answers to outstanding problems in mathematics and physics.

---

[8]https://commons.wikimedia.org/wiki/File:Sir_Isaac_Newton._Mezzotint_by_J._Faber,_junior,_1726, _after_Wellcome_V0004265.jpg

3.3. Newton. *Isaac Newton* (1642-1727), whose development of the Calculus (but little of its publication) predates that of Leibniz by a decade or so, was more cautious in his description of infinitesimals.

To justify calculations that included infinitesimals he therefore relied primarily on his physical intuition and on 'motion'. He discussed the distance covered by a moving 'particle' tracing out a curve over an infinitesimal time period. He described the 'flow', i.e. the change in position, of a variable x (the *fluent*) over an infinitesimal 'instant' o by means of its velocity or *fluxion*, $\dot{x}$. The change in position is then provided by the product $\dot{x}o$. Similarly, for a variable y, whose values depend on those of x, the change in position is $\dot{y}o$. For example, if $y = x^2$, this yields

$$\dot{y}o = (x + \dot{x}o)^2 - x^2 = 2x\dot{x}o + (\dot{x}o)^2.$$

The relative velocity is the *ratio* of the two changes in position,

$$\frac{\dot{y}o}{\dot{x}o} = 2x + \dot{x}o.$$

Now Newton argues that o is infinitesimal, so the final term can be neglected, and the ratio of the two fluxions is therefore given by $2x$. For each x in the abscissa (the x-axis), this ratio is then interpreted as the *tangent* to the curve $y = x^2$ at the point x, and measures the curve's *instantaneous rate of change* at this point. He uses this approach to analyse a wide range of curves. Clearly, unless the curve is a straight line, the slope of the tangent will vary as x varies.

Similarly, Newton computed the fluxion of $y = x^3$ by considering

$$\dot{y}o = (x + \dot{x}o)^3 - x^3 = 3x^2\dot{x}o) + 3x(\dot{x}o)^2 + (\dot{x}o)^3,$$

so that the ratio in the changes of position becomes

$$\frac{\dot{y}o}{\dot{x}o} = 3x^2 + 3x(\dot{x}o) + (\dot{x}o)^2,$$

which he equates with $3x^2$ as the last two terms again 'vanish'.

To handle general integral powers $y = x^n$, Newton used the *binomial theorem*, which (as above for $n = 2, 3$) expresses $(a + b)^n$ as a finite sum of terms in $a^k b^{n-k}$ for $k = 0, 1, 2, ...n$. The *binomial coefficient* of $a^k b^{n-k}$ takes the form $\binom{n}{k} = \frac{n(n-1)...(n-k+1)}{k!}$, where the denominator (k *factorial*) is given by the product $k! = 1 \times 2 \times 3 \times ... \times k$. These coefficients can be read off from the rows of the famous 'triangle' of *Blaise Pascal* (1623-1662) shown in Figure 30, where each term is the sum of the two diagonally above it.

Newton applied the binomial theorem to find the fluxions of $y = x^n$ just as in the examples we computed above, neglecting all terms that still include o after division by $\dot{x}o$. This ensures that the ratio of the fluxions (the relative rate of change in position) is $\frac{\dot{y}}{\dot{x}} = nx^{n-1}$. This represents the *instantaneous*

$$
\begin{matrix}
 & & & & & & & 1 & & & & & & & \\
 & & & & & & 1 & & 1 & & & & & & \\
 & & & & & 1 & & 2 & & 1 & & & & & \\
 & & & & 1 & & 3 & & 3 & & 1 & & & & \\
 & & & 1 & & 4 & & 6 & & 4 & & 1 & & & \\
 & & 1 & & 5 & & 10 & & 10 & & 5 & & 1 & & \\
 & 1 & & 6 & & 15 & & 20 & & 15 & & 6 & & 1 & \\
1 & & 7 & & 21 & & 35 & & 35 & & 21 & & 7 & & 1
\end{matrix}
$$

Figure 30. Pascal's Triangle

rate of change in the variable $y = x^n$ as x varies. Newton assumed (without proof) that the fluxion of the polynomial $y = a_n x^n + a_{n-1} x^{n-1} + ... + a_1 x + a_0$ can then be found term-by-term.

The method of fluxions was of critical importance in Newton's physics, where, for a moving body, velocity (the rate of change in position) and acceleration (the rate of change in velocity) are calculated as fluxions, so that acceleration is the second-order fluxion of position (or distance travelled).[9] Newton's second law $F = ma$ (force equals mass times acceleration) makes use of a as the second fluxion of position. Similarly one may repeat the process to seek the kth fluxion for any whole number k. In modern terminology this is called the kth *derivative*, written by Leibniz as $\frac{d^k y}{dx^k}$.

Finding the fluxion enabled Newton to compare the differences in position (the increments) of the x and y variables (hence *differentiation*). Going in the opposite direction turned out to be the same as Leibniz' summation, i.e. *integration*. This required Newton to find the fluent y when its fluxion is known. He was clearly aware of the inverse relationship between these two processes. He explained this by example in his *De Analysi* (written in 1669 but not published until 1711).[10] In his example he assumed the area under a certain unknown curve, taken from the origin up to some unspecified value x_0, to be given as $\frac{2}{3} x_0^{\frac{3}{2}}$. He then reversed his perspective: considering $z = \frac{2}{3} x^{\frac{3}{2}}$ as the fluent, he showed that its fluxion is $y = x^{\frac{1}{2}}$, and that this is the curve for which the area under the graph was given as $\frac{2}{3} x_0^{\frac{3}{2}}$. This demonstrated by example now the two operations may be seen as inverses of each other (see Figure 31).

---

[9]Velocity is the fluxion (rate of change) of distance, while acceleration is the fluxion of velocity, so that acceleration is the second-order fluxion of distance.

[10]His method of fluxions was expounded in some detail in his Latin treatise *Methodus fluxionem et serium infiniorum*, written in 1671. This important, but technically difficult, work did not find a publisher in Newton's lifetime, and first appeared in print in 1736 (in an English translation) and then in 1744 (retranslated) in Latin.

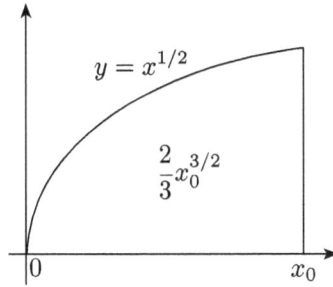

Figure 31. Newton's calculus

Newton extended his rules for finding fluxions to curves given by *fractional* powers of x, such as $x^{\frac{3}{2}}$. He showed, more generally, that the area under the graph of $y = x^{\frac{m}{n}}$ is given by

$$z = \left(\frac{m}{m+n}\right)x^{\frac{m+n}{n}}.$$

This was, in itself, a remarkable extension of the known results involving fluxions at that time. To achieve this, Newton extended the binomial theorem to *fractional powers*. This meant that $(a+b)^{\frac{m}{n}}$ could be expanded as an *infinite series* (a series with infinitely many terms), whose n^{th} term took the form $a_n x^n$, where Newton found each constant coefficient a_n by interpolation and analogy, rather than by a formal proof.[11]

Examples like this led him to assert confidently in his 1669 treatise *De Analysi*:

And whatever common analysis performs by equations made up of a finite number of terms (whenever it may be possible), this method may always perform by infinite equations: in consequence, I have never hesitated to bestow on it also the name of analysis.

Newton employed series expansions of functions to great effect, enabling him to build extensive tables of fluxions (what we call *derivatives*) derived from given fluents. Conversely, he obtained the fluents from given fluxions (our *integrals* or *antiderivatives*), for a wide range of curves, always working term by term from the series expansion. His success encouraged others to work with series containing infinitely many terms containing increasing powers of x, in the same way as they had done with polynomials. They felt that the *'power series'* so obtained could now safely be regarded as obeying the same algebraic laws as polynomials and provided an alternative representation of a host of different curves. Wallis (whose interpolation methods Newton had used extensively) expressed this conviction most clearly in his

---

[11]More details, including a proof of the binomial theorem, can be found in *MM*.

Algebra, arguing that infinite series *'intimate the designation of some particular quantity by a regular Progression or rank of quantities, continually approaching to it; and which, if infinitely continued, must be equal to it'.*

Concerns over the validity of the methods employed in the Calculus took some time to emerge, perhaps obscured by the evident success of Newton's and Leibniz' results in solving outstanding problems. A more immediate question concerned the meaning of infinitesimals in the conceptual frameworks that Cavalieri, Leibniz and Newton had employed. Some of these concerns will be outlined in Section 4.

Insistence on motion remains present in many of Newton's mathematical writings, and he uses it in his attempt to justify his fluxional calculus. He argues that mathematical quantities are not to be seen as *'composed of Parts extreamly small, but as generated by a continual motion'.*

In his famous *Principia Mathematica* (1687) he states: *'Quantities, and the ratios of quantities, which in any finite time converge continually to equality, and before the end of time approach nearer to each other than by any given difference D, become ultimately equal.'* (This is, in effect, his version of the Principle of Continuity.)

Finally, in *Tractatus de quadratura curvarum* (1693) he argues that *'fluxions are very nearly the Augments of the Fluents, generated in equal, but infinitely small parts of Time, and to speak exactly, are in the **Prime Ratio** of the nascent Augments.....'Tis the same thing if the Fluxions be taken in the **Ultimate Ratio** of the Evanescent Parts'.*

However, recourse to infinitely divisible *time,* rather than space, does not offer a way out of the dilemma. Newton's calculations clearly violate another key principle that he had stated as: *'...Errours, tho' never so small, are not to be neglected in Mathematicks'.* His instant o cannot be used when dividing the 'instantaneous' change in y by the corresponding change in x and then be 'neglected' in the same breath!

3.4. Euler and the natural logarithm. Nevertheless, development of the methods of the Calculus and its many applications led to an increasingly sophisticated quantitative analysis of all forms of dynamics throughout the eighteenth century, with most practitioners paying scant attention to the underlying inconsistencies in its mathematical foundations. The prolific and highly influential Swiss mathematician *Leonhard Euler* (1707-1783), writing in the 1740s, made clear that he regarded the use of infinitesimals as the basic tool in handling differentiation, describing it as *'a method for determining the ratios of the vanishing increments that any functions take on when the variable, of which they are functions, is given a vanishing increment'.*[12]

---

[12]Euler had severe problems with his eyesight from at least 1738, and became completely blind in the 1760s. Despite this handicap, he produced over 800 books and articles, ranging over

An important innovation in his writings was that the concept of *function* (usually given by a formula) replaced that of a *curve* as the basis of his analysis – thereby moving it away from visual representations while greatly widening its scope. In his ground-breaking 1748 treatise *Introductio in analysin infinitorum* (Introduction to analysis of the infinite) Euler says: *'A function of a variable quantity is an analytic expression composed in any way whatsoever of the variable quantity and numbers or constant quantities.'* Although this definition was later modified, it marked a significant shift away from reliance on curves and geometric representation.[13]

In his *Introductio* he displayed great (and only occasionally unfounded) confidence in dealing with equations with infinitely many terms, and with power series in particular. He also used *infinitesimals*, as well as their reciprocals, *infinite* 'numbers', freely. His genius lay in (usually) arriving at correct results, even though the methods he used often could not be justified rigorously.

In particular, Euler set out to give definitions of key classes of function, such as polynomials, exponentials (which for him were *'simply powers whose exponents are variable'*) and logarithms (which were the *'inverse of these'*). His derivation of the *natural logarithm* provides a good case study in mid-eighteenth century Calculus.

John Napier, when developing his logarithmic tables (see **Chapter 3**), had correctly appreciated the usefulness of using a geometric progression with common ratio $\frac{s_{n+1}}{s_n}$ very close to 1, but unfortunately insisted on using a *decreasing* sequence, as he wished to keep his 'whole sine' sufficiently large.

The proof that, as n grows ever larger, the numbers $s_n = (1 + \frac{1}{n})^n$ will in fact settle down to a definite value (lying between 2 and 3) is credited to Jacob Bernoulli, who had been investigating growth rates of investments accruing at compound interest rates, with compounding happening at ever shorter time intervals. At that stage, however, no-one had yet associated the limiting value with logarithms.

It was Euler who first considered the elusive 'limit' e of the increasing sequence $(s_n)_{n\geq1}$ with $s_n = (1+\frac{1}{n})^n$ as the base of a system of logarithms. In Chapter 6 of his *Introductio*, Euler discussed exponents and logarithms and also applied his results to problems such as population growth rates and the amortisation period of a loan attracting periodic compound interest.

In Chapter 7 he considered logarithms relative to a general base $a > 1$, and noted that, if ω is an 'infinitely small quantity', he can write $a^\omega = 1 + \psi$, where ψ is also infinitely small. He assumed that his choices would allow

---

almost all the areas of mathematics of his time, and introducing several new areas of research. In his later years, his writings were dictated to his sons.

[13]The development of the modern concept of function is outlined in *MM*.

Figure 32. Leonhard Euler by J.F.A. Darbes, 1778.[14]

him to write $\psi = k\omega$ for some finite k, so that $a^\omega = 1 + k\omega$, hence

$$\omega = \log_a(1 + k\omega).$$

Taking an 'infinite number' j, he then expressed $a^{\omega j} = (a^\omega)^j = (1 + k\omega)^j$ as a power series.

To do this, Euler applied the binomial theorem directly to the sum of 1 and the infinitely small number $k\omega$, as well as using the infinite power j, writing

$$(1 + k\omega)^j = 1 + \frac{j}{1!}k\omega + \frac{j(j-1)}{2!}k^2\omega^2 + \frac{j(j-1)(j-2)}{3!}k^3\omega^3 + \dots$$

Next, he again supposed that $j = \frac{z}{\omega}$ for some *finite* z, so that $z = \omega j$ and $\omega = \frac{z}{j}$. Since $a^z = (a^w)^j = (1 + k\omega)^j$, the series expansion now read

$$a^z = 1 + \frac{1}{1!}kz + \frac{j(j-1)}{2!}k^2(\frac{z}{j})^2 + \frac{j(j-1)(j-2)}{3!}k^3(\frac{z}{j})^3 + \dots$$

$$= 1 + \frac{1}{1!}kz + (\frac{j-1}{j})\frac{1}{2!}k^2z^2 + (\frac{(j-1)(j-2)}{j^2})\frac{1}{3!}k^3z^3 + \dots$$

But since j is infinite, he argued that the ratios $\frac{j-1}{j}$, $\frac{(j-1)(j-2)}{j^2}$, etc., would all cancel (!!), leaving him with the expansion

$$a^z = 1 + \frac{kz}{1!} + \frac{k^2z^2}{2!} + \frac{k^3z^3}{3!} + \dots$$

with all three of a, z, k as *finite* numbers.

[14]https://commons.wikimedia.org/wiki/File:Leonhard_Euler_by_Darbes.jpg

The simplest case was $k = 1$. Euler now reserved the symbol e for the value of a in that case, deriving a formula that has become a staple of modern mathematics:

$$e^z = 1 + \frac{z}{1!} + \frac{z^2}{2!} + \frac{z^3}{3!} + \ldots.$$

Having taken $k = 1$, the relation $a^{wj} = (1 + kw)^j$ that he started with now read $e^z = (1 + w)^j = (1 + \frac{z}{j})^j$ for this 'infinite' value of j. Taking $z = 1$, he treated $(1 + \frac{1}{j})^j$ as the *limiting value* of the sequence $s_n = (1 + \frac{1}{n})^n$ when n grows indefinitely. He had therefore found the limiting value (as n grows) of the s_n to be the number

$$e = 1 + \frac{1}{1!} + \frac{1}{2!} + \frac{1}{3!} + \ldots.$$

Euler proved that e is *irrational*.   He calculated its decimal expansion to 23 decimal places, and proceded to use e as the base for what we now call *natural logarithms*.[15] When $x = e^y$, we write $y = \log_e x$.

In order to approximate the natural logarithm of a given positive number x, Euler used calculations similar to the above (again using infinitesimal and infinite numbers freely) to derive infinite series expansions for $\log_e(1 + x)$ and $\log_e(1 - x)$, arriving at the following series, from whose partial sums such logarithmic tables could be established:

$$\log_e(\frac{1 + x}{1 - x}) = 2(x + \frac{x^3}{3} + \frac{x^5}{5} + \ldots + \frac{x^{2n-1}}{2n - 1} + \ldots)$$

He was not yet done. Leibniz and Johann Bernoulli had expressed conflicting views on the nature of $\log_e(-x)$ and their discussion led Euler to consider how to extend the logarithmic function to negative numbers. Bernoulli had argued that $\log_e(-x)$ should equal $\log_e(x)$, since both yield the derivative $\frac{1}{x}$; while Leibniz argued that the rule $\frac{d}{dx}(\log_e x) = \frac{1}{x}$ assumed that $x > 0$. Euler pointed out that two functions that differ by a constant have equal derivatives, so that one cannot conclude that the functions themselves will be equal if their derivatives are equal.

By definition of the logarithm, $\log_e(-x) = \log_e((-1) \times x) = \log_e(x) + \log_e(-1)$, as Euler pointed out. To determine the value of the final term on the right, he would use the familiar *de Moivre formulae*, which, in his hands, became a fundamental tool in complex analysis.[16] Writing i for the 'imaginary unit' $\sqrt{-1}$ (this became the standard notation) he knew that, for any

---

[15]The reason for calling logarithms to this base *natural* relates to its definition, which can be given (as above) as the inverse of the exponential function, or as the integral of the function $g(x) = \frac{1}{x}$. The exponential function $f(x) = e^x$ has the unique property that it equals its derivative: $f'(x) = e^x$, so that, at any point x, its instanteous *rate of growth* is equal to its value. This concept (and its generalisations) has many applications to models of population growth, continuous compounding, etc. See [6], and see *MM* for a proof that the number e is irrational.

[16]See *MM* for the derivation of these formulae.

$n \geq 1$,

$$\cos(n\theta) + i\sin(n\theta) = (\cos\theta + i\sin\theta)^n, \quad \cos(n\theta) - i\sin(n\theta) = (\cos\theta - i\sin\theta)^n.$$

Adding provides the identity $\cos(n\theta) = \frac{(\cos\theta + i\sin\theta)^n + (\cos\theta - i\sin\theta)^n}{2}$, and, setting $x = n\theta$ and taking n 'infinite' (so that $\theta = \frac{x}{n}$ is infinitesimal), Euler deduced that $\cos(\frac{x}{n}) = 1$ and $\sin(\frac{x}{n}) = \frac{x}{n}$. Substituting this into the above he obtained,

$$\cos x = \frac{(1 + i\frac{x}{n})^n + (1 - i\frac{x}{n})^n}{2} = \frac{e^{ix} + e^{-ix}}{2},$$

where the final identity follows because, when n is infinite and z finite, we have $e^z = (1 + \frac{z}{n})^n$ as noted above, and can apply this to the finite (imaginary) quantity $z = ix$. An exactly analogous argument with $i\sin(n\theta)$ shows that $i\sin x = \frac{e^{ix} - e^{-ix}}{2}$, so that he derived what we now call *Euler's identity*:

$$e^{ix} = \cos x + i\sin x.$$

He had shown that complex exponentials could, as he put it, *'be expressed by real sines and cosines'*.

A celebrated identity arises when we take $x = \pi$: we have $e^{i\pi} = -1$, i.e.

$$e^{i\pi} + 1 = 0.$$

This, it is somtimes argued, links the five 'most important' numbers in mathematics: $0, 1, e, \pi$ and i. It may also have served to persuade many observers that the 'mysterious' square root of -1 needed to be understood more fully.

Finally, taking the natural logarithm on both sides of the identity $e^{i\pi} = -1$, Euler noted that $\log_e(-1) = i\pi$. Thus the logarithm of a negative number is purely imaginary. Euler went on to deduce (correctly), that the logarithm of a complex number is not single-valued, but has *infinitely many branches* – but we will leave the matter there.

4. Critique of the calculus

In Britain, serious questions about the foundations of the Calculus were raised publicly soon after Newton's death in 1727, in a way that could not easily be ignored. The fundamental inconsistency of early Calculus techniques was seized upon by the philosopher and cleric *George Berkeley* (1685-1753), Bishop of Cloyne in Ireland. Berkeley's explicit purpose was to defend religious faith against assertions by the Astronomer Royal *Edmund Halley* and others (although not Newton himself) that scientific and mathematical progress had rendered faith in scriptural revelation redundant.[17] In his 1734 tract *The Analyst* (whose subtitle begins: *A DISCOURSE addressed to an infidel MATHEMATICIAN...*) Berkeley examines whether the new Calculus really was as soundly based as had been claimed; or as he put it:

---

[17]Halley had mocked the earlier tract *Alciphron* by Berkeley; it is also claimed that Halley had persuaded a friend of Berkeley's to renounce religion on his deathbed.

Whether such mathematicians as cry out against mysteries have ever examined their own principles?

(The Analyst, Question 63)

Berkeley argued, correctly, that the Calculus of Newton and Leibniz rested on the use of infinitely small quantities whose existence was unproven and logically dubious. He pointed out that such infinitesimals were treated as actual (non-zero) quantities in calculations, yet were later declared to be negligible. Fastening on Newton's notion of ultimate ratios, he asked memorably:

"And what are these Fluxions? The Velocities of evanescent Increments? And what are these same evanescent Increments? They are neither finite Quantities nor Quantities infinitely small, nor yet nothing. May we not call them the ghosts of departed quantities?"

His critique, arguing that mathematicians are as reliant on faith as theologians, and worked by *'submitting to authority, taking things on trust'*, hit home in British mathematical circles, and provided motivation for various attempts by prominent mathematicians to improve the foundations of the Calculus.[18] Of these responses perhaps the most complete was *Treatise on fluxions* (1742) by the Scottish mathematician *Colin Maclaurin* (1698-1746), a major two-volume work which set the Calculus in a geometric framework and further developed the theory of power series, but did not really rebut Berkeley's critique.

Nonetheless, as we have seen, and just as proved to be the case with the discovery of incommensurables like $\sqrt{2}$ more than two millennia earlier, a lack of proper foundations for their new methods of analysis in no way delayed most mathematicians (especially on the Continent) in their development of the Calculus and its many applications in the natural sciences throughout the seventeenth and eighteenth centuries.

It continued to cause concern to philosophers, however, and even troubled the eminent empiricist *David Hume* (1711-1776). On the one hand, Hume famously ends his *Enquiry into Human Understanding* [22] with a clarion call to his readers:

If we take in our hand any volume; of divinity or school metaphysics, for instance; let us ask; **Does it contain any abstract reasoning concerning quantity or number?** *No.* **Does it contain any experimental reasoning concerning matter of fact and existence?** *No. Commit it then to the flames; for it can contain nothing but sophistry and illusion.*

However, despite his high regard for mathematical reasoning, just a few pages earlier in the same volume Hume expresses with great clarity

---

[18]Berkeley also questioned whether irrationals, such as the diagonal of the unit square, should be treated as numbers. This aroused rather less concern at the time.

his acute anxiety over the meaning of an apparent hierarchy of infinitely divisible quantities:

The chief objection against all **abstract** *reasonings is derived from the ideas of space and time; ideas, which, in common life and to a careless view, are very clear and intelligible, but when they pass through the scrutiny of the profound sciences (and they are the chief object of these sciences) afford principles, which seem full of absurdity and contradiction. No priestly* **dogmas***, invented on purpose to tame and subdue the rebellious reason of mankind, ever shocked common sense more than the doctrine of the infinite divisibility of extension, with its consequences; as they are pompously displayed by all geometricians and metaphysicians, with a kind of triumph and exultation. A real quantity, infinitely less than any finite quantity, containing quantities infinitely less than itself, and so on* **in infinitum***; this is an edifice so bold and prodigious, that it is too weighty for any pretended demonstration to support, because it shocks the clearest and most natural principles of human reason.*

Hume's particular example, which, as he points out, relies only on *'the clearest and most natural'* chain of reasoning (properties of circles and triangles), is the notion of the angle of contact, or *horn angle*, between a straight line and a curve (such as that between the circumference of a circle and its tangent). This angle is shown to be *'infinitely less than any rectilineal angle'* and can be made ever smaller simply by increasing the diameter of the circle. While the proof of this claim seems *'as unexceptionable as that which proves the three angles of a triangle to be equal to two right angles'*, it palpably offends common sense, in his view. Hume could equally well have said the same about perceptions that the number line can contain similarly incomparable quantities.

CHAPTER 6

From Calculus to Analysis

The understanding must not therefore be supplied with wings, but rather hung with weights, to keep it from leaping and flying. Now this has never yet been done; when it is done, we may entertain better hopes of science.

Sir Francis Bacon *Novum Organum*, 1620

Summary

Early in the nineteenth century attention turned decisively to the urgent need to underpin the highly successful techniques of the Calculus with a logically consistent conceptual framework. Most historians agree that the prolific French mathematician *Augustin-Louis Cauchy* (1789-1857) was a key figure in this development. Cauchy used the *limiting value* of a variable quantity as the cornerstone of his theory. This led him to a more rigorous analysis of the operations of the Calculus, even though his celebrated text *Cours d'Analyse* continued to employ *infinitesimals* as 'indispensible' tools.

Cauchy's definitions set the scene for putting the theorems of Calculus on a sounder footing. His formulations were eventually superseded: much of what we now call *real analysis* stems from the pioneering work of the influential *Karl Weierstrass* (1815-1897), further elaborated by his pupils and contemporaries in Prussia. This work was only completed definitively between 1850 and 1870.

In this chapter our focus is on the ideas introduced by Cauchy that, together with the work of his successors, would highlight the need for a new definition of the underlying number system.

1. D'Alembert and Lagrange

Bishop Berkeley's critique of the Calculus, while influential in drawing responses from Newton's followers, seems to have had less traction with Continental mathematicians such as Euler and the Bernoullis, whose conceptions and use of infinitesimals also attracted critical comment, notably

   https://doi.org/10.11647/OBP.0236.06

from the Dutch philosopher *Bernard Nieuwentijt* (1654-1718). As the eighteenth century drew to a close, however, the task of reconciling the un-
doubted success of infinitesimal techniques in the Calculus with their logical ambiguity and lack of clear definition was increasingly recognised, although a convincing resolution of the issues involved would take considerable time.

In contrast to Euler's confident use of the infinitely small (and large),
his near contemporary *Jean le Rond d'Alembert* (1717-1783) argued, implicitly
following Newton, that defining the derivative $\frac{dy}{dx}$ as the *limit* of the ratios
of two quantities was the *'neatest and most precise'* definition of this concept.
D'Alembert's definition appeared in the influential *Encyclopédie ou Diction-
naire Raisonneé des Sciences, des Artes et des Métiers* (9th edition, 1765) which
he co-edited with *Diderot*.

His definition of the concept of limit was close to Newton's:

*One magnitude is said to be the limit of another magnitude when the second
may approach the first within any given magnitude, however small, though the second magnitude may never exceed the magnitude it approaches.*[1]

Since he insisted on the limit being approached from below, this formulation was less complete than the definition of limits used today. Although
D'Alembert does not make it clear, the implication is that the second magnitude is *variable*, taking on values that will eventually become arbitrarily close
to the fixed value of the first magnitude. There is then no need to insist, as
he does, that none of these values are larger than the limit.

He continued:

Strictly speaking, the limit never coincides [with]*, or never becomes equal to,
the quantity of which it is the limit; but the latter approaches it more and more and
can differ from it as little as one wishes.*

Even though his formulation does not quantify clearly what 'can differ
from it as little as one wishes' means, his statement may be an attempt to
defuse criticisms such as Berkeley's *'ghosts'* comment, by arguing that no
division by zero actually takes place in the Calculus. In an article on the
differential he was even stronger in his rejection of infinitesimals:

*the differential calculus does not necessarily assume the existence of these quan-
tities...We shall say that there are no infinitely small quantities in the differential
calculus.*

D'Alembert's comments were premature in one important respect: the
details of operating consistently with limits had yet to be worked out. This

---

[1]Note the use of *magnitude* throughout this comment, which suggests that d'Alembert
considers the quantities as varying continuously, in line with Aristotle's distinction between
continuous (geometrical) 'magnitudes' (as used in Greek 'method of exhaustion' proofs) and
discrete (arithmetical) 'multitudes'.

meant that he did not provide a workable definition of limit on which to build the edifice of the Calculus. It is in this sense that his comments may be regarded as articulating a future research programme rather than a fully worked out method of justifying the operations of the Calculus.

From 1787 until his death, *Joseph-Louis Lagrange* (1736-1813) was one of the major figures in the vibrant mathematical scene that had developed, centered on the *Academie Francaise,* in Paris during the eighteenth century. Lagrange was born in Turin, but changed his Italian name (Giuseppe Lodovico Lagrangia) after an early career in Turin, followed by 20 years in Berlin. He became a member of the *Academie,* where his influential *Mecanique Analytique* (written while in Berlin) was published the following year. Notable for its treatment of differential equations, this summarised all the work done on mechanics since Newton, transforming it into a treatise on mathematical analysis, with the emphatic statement:

'*One will not find figures in this work. The methods that I expound require neither constructions, nor geometrical or mechanical arguments, but only algebraic operations, subject to a regular and uniform course*'.

Appointed as the first professor in analysis at the new École Polytechnique in 1794, he published *Theorie des fonctions analytiques* in 1797, making his objective an exposition of

'*... the principles of the differential calculus, freed from all consideration of the infinitely small or vanishing quantities, of limits or fluxions, and reduced to the algebraic analysis of finite quantities'*,

since, in his view,

'*...the ordinary operations of algebra suffice to resolve problems in the theory of curves*'.

In contrast to d'Alembert's comments, Lagrange's formulations did not include any concept of limit. Moreover, he was critical of earlier attempts using infinitesimals and geometric intuition as the basis of the theory. He sought, instead, to base the Calculus on purely algebraic concepts. In a paper in 1772 he had attempted to define the *derivative* (Newton's fluxion) in terms of (infinite) *Taylor series,* whose first $k+1$ terms constitute the kth Taylor polynomial.

This polynomial, named for *Brook Taylor* (1685-1741), a contemporary of Newton, is expressed in powers of $(x - a)$ and serves to approximate a given function f at points x near a fixed point a. (Recall that in Taylor's time the function f was regarded as describing a curve $y = f(x)$ in the Cartesian plane.) For each $k \geq 0$, the coefficient of $(x - a)^k$ is given as the value of the

k^{th} derivative (or *derived function*) $f^{(k)}$ at a, divided by $k! = 1 \times 2 \times 3 \times ... \times k$:[2]

$$T_k^f(x) = f(a) + \frac{1}{1!}f'(a)(x-a) + \frac{1}{2!}f''(a)(x-a)^2 + ... + \frac{1}{k!}f^{(k)}(a)(x-a)^k.$$

Lagrange recognised that if, for all x in an interval around a, the *remainder term* $R_{k+1}(x) = f(x) - T_k^f(x)$ can be made arbitrarily small by choosing k large enough, the resulting infinite series would *represent* the function f on that interval.[3]

On the *assumption* that every function f to be considered in mathematical analysis has a power series representation of the form

$$f(x) = a_0 + a_1(x-a) + a_2(x-a)^2 + ... + a_k(x-a)^k + ...$$

for some sequence of coefficients $(a_k)_{k \geq 0}$, Lagrange now argued that he could *define* the kth derivative of f at a simply as the product $k!a_k$, where a_k is given in the power series expansion. Hence derivatives would be defined purely algebraically, and thus be redeemed from '*all consideration of the infinitely small, or vanishing quantities, of limits and from fluxions...*' .

Lagrange's terminology and his emphasis (following Newton) on the utility of power series representations were widely accepted. However, his optimistic claim that the derivative could be *defined* purely algebraically did not find quite the same resonance with other major figures at the time. Even if his assumption had proved correct, the question would remain how the coefficients of the power series were to be found—if not by means of Taylor series, which would, in turn, have to be computed by first calculating the derivative of each power of x. In any event, Cauchy was to show that Lagrange's assumption was unjustified. The essential problem is that even relatively 'well-behaved' functions, with derivatives of all orders throughout an interval, need *not* necessarily be represented there by their Taylor series.

Despite his protestations, Lagrange himself may have been less than wholly convinced that he had banished infinitesimals from the Calculus. While he was still director of the mathematical section at the Berlin Academy, a prize problem was issued in 1784 for '*a clear and precise theory of what is called the infinite in mathematics*'. In 1786 the prize was awarded to *Simon Lhuilier* (1750-1840) for a rather turgid study of limits in the style of d'Alembert, which did not really settle the issues it addressed and was bedevilled by confusing terminology. Nonetheless, in a 1795 revision of his essay, Lhuilier made the important—if belated—observation that a limit may

---

[2]This terminology and notation for the kth derivative, widely used today, was introduced by Lagrange.

[3]Here, and below, for any two real numbers $c < d$ an *interval* with these two endpoints contains all real numbers lying between c and d. The *open* interval (c, d) consists of the numbers x satisfying $c < x < d$, while the *closed* interval also includes the endpoints. (For the present, we still interpret 'real numbers' as points on the 'geometric number line'.)

be approached by an oscillating sequence of values, not merely from above or below, thus widening d'Alembert's definition.

Concern about the confused state of 'analysis', as the elaboration and application of the Calculus had become known, led to various Academies offering prizes for work to clarify these matters. The 'research programme' implicit in d'Alembert's *Encyclopédie* entry clearly still had some way to go. An influential, but by no means conclusive, contribution was made by *Lazare Carnot* (1753-1823), a mathematician and military engineer who was also prominent as a politician in the French Revolution and under Napoleon. In two editions of his widely read *Reflections* (1797, 1813), the latter of which proved especially popular, he surveyed several different attempts to resolve the conceptual questions that had been raised, although Lagrange's approach did not feature in this work. Carnot favoured the use of infinitesimals, whose definition he sought to clarify by means of limits. Yet, unlike Cauchy, less than a decade later, he appeared unable to provide a workable definition of the elusive concept of *limit*.

A refutation of Lagrange's over-optimistic claim that, instead of using infinitesimals, one can always define the derivative algebraically by means of Taylor series, appeared nearly a decade after his death. In an 1822 paper, Cauchy proved that the function f given by $f(x) = e^{-\frac{1}{x^2}}$ when $x \neq 0$, and $f(0) = 0$, has derivatives of all orders throughout (even at 0), yet cannot be represented by its Taylor series (expanded at $a = 0$) at any point other than at 0 itself!

In fact, Lagrange had earlier reconciled himself to infinitesimals, at least after a fashion. Near the end of his life the second edition of his masterpiece, *Mecanique analytique*, published in 1811, contained the (perhaps rueful) remark:

When we have grasped the spirit of the infinitesimal method, and have verified the exactness of its results either by the geometrical method of prime and ultimate ratios, or by the analytical method of derived functions, we may employ infinitely small quantities as a sure and valuable means of shortening and simplifying our proofs.

However, the spirit of the age was that such matters needed to be addressed afresh. The philosophers of the Enlightenment placed emphasis on reason and the individual, rather than on tradition. Together with the impact of the American and French Revolutions this galvanised intellectuals all over Europe to question established authority and to examine the foundations of knowledge anew. Meanwhile, the rapidly developing Industrial Revolution provided scientists with new problems for solution, often leading mathematicians to extend and refine the tools of the Calculus. It was increasingly clear that lack of precision in the foundations of the Calculus could inhibit further progress.

During this period the academies that had flourished at various aristo-
cratic courts were gradually being supplanted by universities as the principal centres of learning. This change had been accelerated by the establishment of the advanced *Écoles* (*Normale* and *Polytechnique*) in Paris soon
after the French Revolution. This had further cemented the dominance of
Paris as the acknowledged international centre of excellence in mathematical research, while the increasing emphasis on the formal instruction of students led to more systematic attempts finally to provide the Calculus with
the sound foundations that earlier critics had called for in vain.[4]

2. Cauchy's 'Cours d'Analyse'

Today, mathematicians accept the Calculus as grounded solidly and ap-
propriately in the real number system.[5] This defines the *continuum* on which
functions studied in basic Calculus are defined, and provides the spring-
board for the complex numbers and generalisations to spaces of many (even
infinitely many) dimensions. Popular science texts, while often applying the
methods of 'school calculus', seldom address the underlying concepts, pre-
ferring (as we have done so far) to rely on geometric intutition, representing
real numbers by 'points' on a 'number line'.

Not surprisingly, mathematics undergraduates are initially taken aback
when meeting the definitions and proofs required to justify the basic operational techniques of Calculus in more detail—especially the need for a rigorous definition of the underlying number system—and frequently profess
themselves bewildered by the apparently needless complexity of the subject. This becomes less surprising if we reflect that it took the mathematical
world fully two centuries to arrive at a (more or less) universal consensus
on these matters.

Augustin-Louis Cauchy entered the École as a student aged 16, in 1805.
He studied analysis under *Sylvestre Francois Lacroix* (1765-1843), who employed a geometrically based limit concept in the style of d'Alembert's programme. After the fall of Napoleon Bonaparte, and amid political upheavals
during the restoration of the Bourbon monarchy in 1814, the eminent scien-
tist *Pierre Simon Laplace* (1749-1827) was put in charge of the reorganisation
and re-opening of the École in 1816. Cauchy, then aged 27, replaced *Louis
Poinsot* (1777-1859) as professor of analysis.

Cauchy's immediate predecessors had seen little point in addressing
questions of logical rigour when teaching mathematics to engineers, which

---

[4]The dominance of Paris was a significant factor spurring the development of a rigorous
approach to university tuition in Prussia, spearheaded by Wilhelm and Alexander von Hum-
boldt; so that by 1830, Berlin and Göttingen (where *Gauss* spent the bulk of his career) were to
rival the reputation of Paris.

[5]Arithmetical representations of this number system, dating from 1872, will be considered
in **Chapter 7**.

they regarded as the École's principal objective. In 1810 the analysis course, until then based on limit concepts developed by *Lacroix*, was replaced by one closer to the style of Lagrange and Euler, where algebraic formulae were applied to infinitesimal and infinite quantities in exactly the same way as to finite quantities.

In his highly influential 1821 treatise *Cours d'Analyse*, based on his lectures at the *École Polytechnique* from 1816, Cauchy makes clear that he had decided to employ a very different approach. Today, many historians regard this text as the origin of *mathematical analysis*—by which they mean the study of (real and complex) variables and functions, especially as used in the Calculus.

Cauchy's opponents at the École were fiercely critical of the youthful professor's emphasis on rigour in his teaching and of what they regarded as his resulting neglect of the applications of the Calculus. It has been suggested that Cauchy may at one point have been compelled by the authorities to revert to 'more traditional' methods in his teaching.

In his writings, however, Cauchy is adamant. The introduction to the *Cours* makes plain his objectives:

Regarding methods, I have sought to give them all the rigour one requires in geometry, in such a way as never to resort to reasons drawn from the generality of algebra.

This reference to 'the generality of algebra' is to the algebraic formalism which, in the hands of Lagrange and his predecessors, seemed to Cauchy to have become detached from its roots in Descartes' geometry and had acquired a life of its own, without clearly defining all the objects it described. Cauchy's ambitious aim was to make sense of the new tools that had been invented for the Calculus, employing a degree of rigour similar to that of Euclid's *Elements*. The fact that one could write down an algebraic formula and manipulate it would not suffice to give it meaning, he argued.

In modelling his approach on Euclid, Cauchy was advocating a radical shift in perception that we will examine more closely in the next two chapters. Rather than regarding the objects of mathematical research simply as 'given' – as, for example, in Plato's World of Ideas – and to be manipulated according to various inituitively obvious 'rules', he sought, instead, to provide clearly articulated principles that underlie these rules, as Euclid had done in stating his five postulates.

In practice, Cauchy's analysis did not really meet the standards of clarity set by Euclid's *Elements*, nor did he succeed fully in defining a logically sound basis for the operations of the Calculus. The principal problem was that he continued to rely on the geometric description of the 'number line', on which he based his key concept of *variable quantity*. Thus, for example, the distance between two values that the variable quantity might take was

A^n Cauchy

B^n Augustin Cauchy

Figure 33. Augustin Cauchy by Gregoire et Deneux, 1840.[6]

defined by reference to geometry, rather than given an independent definition. In addition, as we shall see, he continued to rely on the notion of 'infinitesimal' quantities to describe his definitions.

Nonetheless, his *Cours* provides glimpses of a profound philosophical shift that would dominate mathematical research in the nineteenth century and continues today. We should therefore consider his conceptions more closely and trace his influence on later developments.

While Cauchy did not provide an unambiguous definition of the underlying number system in his *Cours,* he clarified the key concepts of his analysis, such as 'continuity' and 'limit' of a function, more precisely than his predecessors had done, and used them to underpin the twin operations of the Calculus, 'derivative' and 'integral'. Yet, despite d'Alembert's optimistic comments half a century earlier, Cauchy (perhaps echoing Lagrange's chastened comment) maintained in the Introduction to his *Cours* that he, too, was *'unable to dispense with'* the use of *'infinitesimal quantities'* in the operational techniques he presented. On the face of it, Cauchy's concept of the continuum therefore appears to include infinitesimals and remains somewhat obscure.

He recognised that, if they were to be useful, infinitesimals had to be defined more precisely. His solution was to subjugate their definition to

───────────────
[6]https://commons.wikimedia.org/wiki/File:Cauchy_Augustin_Louis_dibner_coll_SIL14-C2-03a.jpg

that of his concept of *variable quantity* – which he defined as a quantity which is assumed to *'take on successively several different values, one after the other'* – and thereby to define infinitesimals as variable quantities that eventually 'become infinitely small'.

2.1. Limits of sequences. To achieve this, he gave precedence to the concept of the *limit* of a variable quantity. Cauchy described this verbally:

*When the successively assigned values of the same variable indefinitely approach a fixed value, so that they end up by differing from it by as little as one could wish, the last is called the **limit** of all the others. So, for example, an irrational number is the limit of the various fractions which provide values that approximate it more and more closely...*

He followed this with his definition of infinitesimal quantities:

*...When the successive numerical values of the same variable decrease indefinitely in such a way as to fall below any given number, this variable becomes what one calls an **infinitesimal** or an **infinitely small** quantity. A variable of this kind has zero for its limit.*

Cauchy's infinitesimals thus appear in the guise of variables whose assigned values have limit 0, without actually *attaining* that limit at any stage in a finite number of steps. In that sense his infinitesimals are described by a *process* rather than as actual *quantities*, which in turn makes their use in arithmetic problematic. One drawback is that he never explicitly defines how these values are assigned, although in seeking to reconcile rigour with his *quantitées infiniment petites* he seeks to distinguish them from what we would today regard as real numbers.

Although Cauchy's definition remains silent about the precise nature of the independent variable x, he states that it will take on 'successively assigned values', and demands that these values should 'finally' differ from the limit (which is the 'fixed value' in his definition) 'by as little as one could wish'.

It is perhaps easiest to characterise his definition of the limit of an infinite *sequence*: in that case the sequence of numbers $x_1, x_2, x_3, \ldots$ represents the 'successively assigned values' of a 'variable quantity', numbered by successive natural numbers $1, 2, 3, \ldots$. If, for some fixed number a, the difference between a and x_n can be made smaller than *any* given number $\varepsilon > 0$ by the simple expedient of taking n 'far enough along the sequence' then Cauchy calls a the *limit* of this sequence. Thus: a is the limit of the $(x_n)_n$ provided that the distance between x_n and a will *eventually* become less than any pre-assigned number $\varepsilon > 0$, i.e. for some fixed natural number N, this distance should be smaller than ε whenever $n \geq N$.

We cast this statement in today's terminology:

*The real number a is the **limit** of the sequence of real numbers $(x_n)_{n\geq 1}$ if, given any real number $\varepsilon > 0$, one can find $N \in \mathbb{N}$ such that $|x_n - a| < \varepsilon$ for all $n \geq N$.*

When this requirement is satisfied, we write $a = \lim_{n\to\infty} x_n$, and say that the sequence $(x_n)_{n\geq 1}$ *converges* to a.

Cauchy's definition assumes implicitly that the numbers x_n and a represent points that lie on an underlying number line, rather than defining these numbers arithmetically. But this begs the question how we can compare the distance between x_n and a with the given positive number ε. Thus limits, as defined by Cauchy, cannot be used for a logically coherent *construction* of the continuum as a number system that contains irrationals. To justify his statement (above) that an irrational number a is *'the limit of the various fractions which provide values that approximate it more and more closely'* (here represented by the sequence $(x_n)_n$), Cauchy would have to explain how the approximations are to be computed, and this would presuppose that the irrational had already been defined. This may be evident geometrically, but that is not an arithmetical definition.

Cauchy's attempt to define irrationals via limits may not make sense arithmetically, but his definitions and techniques helped to identify the root of the problem to be addressed, and prepared the ground for its solution, as well as laying the groundwork for modern analysis more generally. One of his most significant innovations was what is now called the *Cauchy criterion* for convergence, and we examine this more closely.

In Chapter 6 of the *Cours d'Analyse* Cauchy defined the *sum* of an infinite series $u_0 + u_1 + u_2 + \dots + u_n + \dots$ (here denoted as $\sum_{n\geq 0} u_n$) as follows:

*Let $s_n = u_0 + u_1 + u_2 + \dots + u_{n-1}$ be the sum of the first n terms, with n designating an arbitrary integer. If, for increasing values of n, the sum s_n approaches indefinitely a certain limit s, the series will be called **convergent**, and the limit in question will be called the **sum** of the series.*

In other words, convergence of the series $\sum_{n\geq 0} u_n$ simply means that the sequence $(s_n)_{n\geq 1}$ of its *partial sums* converges to a finite limit, as defined above. If this sequence does not have a limit, the series $\sum_{n\geq 0} u_n$ is said to be *divergent* and its sum remains undefined.

Cauchy's concern was to provide a clear definition as well as giving a checkable *criterion* for the convergence of series. His criterion states that the convergence of the series $\sum_{m\geq 0} u_m$ is *guaranteed* if, for n sufficiently large, the sums $u_n + \dots + u_{n+k-1}$ can *simultaneously* be made arbitrarily small for every $k \geq 1$.

As he puts it, the sums should, when *'taken, from the first, in whatever number one wishes, finish by constantly having an absolute value less than any assignable limit'*.

Expressed in terms of the partial sums $s_{n+k} = u_0 + \dots + u_n + \dots + u_{n+k-1}$ and $s_n = u_0 + \dots + u_{n-1}$, this requirement is that the sequence of partial sums $(s_n)_n$ should satisfy the criterion given below, where N, n, k as whole numbers.

Given $\varepsilon > 0$, we can find N such that for every $n \geq N$, the following condition holds for all $k \geq 1$:

$$|s_{n+k} - s_n| < \varepsilon.$$

Cauchy stated his criterion as an unproven assertion, an *axiom*. In other words, he asserts that, as long as a series $\sum_{m \geq 0} u_m$ satisfies his criterion, that series will have a finite sum.

We rephrase this criterion more generally in terms of an arbitrary sequence $(x_n)_{n \geq 1}$ of real numbers. The Cauchy criterion then demands that:

given any real number $\varepsilon > 0$, there exists N such that $|x_m - x_n| < \varepsilon$ for all $m, n \geq N$.

Today, such a sequence $(x_n)_n$ is known as a *Cauchy sequence*.

Intuitively, a sequence 'ought to have' a limit provided that 'eventually' any pair x_m, x_n of its members will be 'as close as we wish' to each other. This is what the requirement '$|x_m - x_n| < \varepsilon$ when $m, n \geq N$' ensures. Although Cauchy then *asserts* (without proof) that in these circumstances the limit must exist, the criterion itself makes no mention of the limit.

This fact was central to the model for the real numbers presented by Georg Cantor in 1872, which we discuss in **Chapter 7**. Cantor's model avoids the logical trap that Cauchy fell into. In fact, Cauchy's criterion is one of several versions of the key property (which we call its *completeness*) that distinguishes the real number system from the system of rational numbers. Pictorially we describe this as 'having no gaps'. Cauchy's criterion can fail if we deal exclusively with rational numbers—we had an example of this in **Chapter 1**, featuring a sequence $(r_n)_n$ of rational approximations to $\sqrt{2}$. It can be checked that $(r_n)_n$ is a Cauchy sequence, yet its 'limit' $\sqrt{2}$ is not a rational number. A larger number system, containing no such 'gaps', is required to remedy this defect and ensure that Cauchy's criterion always holds.

Cauchy introduced much of the technical machinery used in modern undergraduate texts on 'Real Analysis' (for example, [27], [40]) especially in his consistent use of *inequalities* to compare various quantities. To illustrate his use of inequalities, we use the most basic example, the well-known *triangle inequality*:

The inequality $|a + b| \leq |a| + |b|$ holds for any (real or complex) numbers a, b.

Geometrically, the triangle inequality is obvious from the Argand diagram in Figure 23(a) (on page 101): in the triangle OQP, the length of the side OP (which represents $|z + w|$) cannot be greater than the sum $|z| + |w|$

of the lengths of the other two sides, OQ and QP. But this argument relies on Euclidean geometry.[7]

With this inequality it is easy to show that any convergent sequence is also a Cauchy sequence, providing the converse to Cauchy's criterion: if $\lim_{n\to\infty} x_n = x$ then for any $\varepsilon > 0$ there exists N such that if $m, n \geq N$, both the inequalities $|x_m - x| < \frac{\varepsilon}{2}$ and $|x - x_n| < \frac{\varepsilon}{2}$ hold. This shows that $(x_n)_n$ is a Cauchy sequence: for $m, n \geq N$,

$$|x_m - x_n| = |(x_m - x) + (x - x_n)| \leq |x_m - x| + |x - x_n| < \varepsilon.$$

3. Continuous functions

In order to tackle, in a consistent manner, the variety of curves to which his predecessors had applied the techniques of the Calculus, Cauchy adopted a definition of the *continuity of a function* that, although given verbally in terms of incremental change, would in practice entail comparing two numerical inequalities.

In the eighteenth century a 'continuous curve' had usually been regarded as as a curve determined by a single 'expression' (or formula). The function concept mostly remained tied to formulae and their visual representation as curves. Continuous curves were seen as those that could be drawn in a single unbroken motion. During a controversy over the initial conditions that should be allowed in the study of a vibrating string, Euler had advocated the idea that the definition of function should be 'completely general', but most practitioners continued to study functions with *contiguous* graphs (drawn in a single motion), or, at worst, ones that consisted of a finite number of pieces.[8]

Cauchy's definition of continuity dispenses with visual images:

The function $f(x)$ will be, between two assigned values of the variable x, a continuous function of this variable if for each value of x between these limits, the value of the difference $f(x + \alpha) - f(x)$ decreases indefinitely with α.

In Cauchy's definition the use of the phrase '*these limits*' simply means that x lies between two '*assigned values*'; i.e. $c < x < d$ for some fixed numbers c, d. In other words, Cauchy considers functions defined throughout an *open interval* $I = (c, d)$, which consists of all points lying strictly between the points c and d. When an interval includes both endpoints, we say that it is *closed* and denote this by $[c, d]$.

The phrase *decreases indefinitely with α* can be understood in the same way as his definition of limit: when successive values are assigned to the variable α (which has limit 0, hence becomes infinitesimal, according to

---

[7]A simple algebraic proof and a summary of key facts about convergent real sequences and series can be found in *MM*.

[8]A summary of the modern function concept can be found in *MM*.

Cauchy's definition) then the absolute values of $f(x + \alpha) - f(x)$ and α decrease to 0 together. In other words, if we change the independent variable x by an infinitesimal amount, the change in $f(x)$ is also infinitesimal. Cauchy did not specify that 'value' means absolute value here—the notation for this came later—but he clearly intended it to be seen as such.

In his proofs Cauchy often linked his definition of continuity to that of limit, making f continuous at a precisely when the function value $f(a)$ equals the limit of the values $f(x)$ when, though *successively assigned values'*, x approaches the point a, We can rephrase this statement as follows:

The function f is continuous at the point a if whenever a sequence (x_n) has limit a, the sequence of values $(f(x_n))_n$ has limit $f(a)$.

This statement defines what is called *sequential continuity* at a point today. For real functions it is logically equivalent to the modern definition of *continuity at a point*, given below.

In 1817, four years before Cauchy's seminal text, the Bohemian mathematician and Catholic priest *Bernhard Bolzano* (1771-1848) developed a formulation of continuity which is essentially the same as that of Cauchy, but without reference to values taken in sequential succession. He proposed to make the somewhat nebulous 'law of continuity' more precise as follows:

If a function $f(x)$ varies according to the law of continuity for all values of x inside or outside certain limits, then, if x is some such value, the difference $f(x + \omega) - f(x)$ can be made smaller than any given quantity provided ω can be taken as small as we please.

This definition does not depend on intermediate concepts such as limit, or limits of sequences. In effect, it relates two inequalities to each other. It avoids any mention of infinitesimals and is very close to the modern definition of continuity at a point (usually attributed to Weierstrass, whose formulation came more than two decades later):

*The function f is **continuous at the point** x if, for given $\varepsilon > 0$ we can find $\delta > 0$ such that $|f(y) - f(x)| < \varepsilon$ whenever $|y - x| < \delta$.*

It is thought unlikely that Cauchy was aware of Bolzano's work when he published the *Cours d'Analyse* in 1821, although it is possible that Lagrange could have been a common source for the need to refine the concept of continuity. Also, while in Berlin in the early 1820s, Niels Abel had learnt of and – according to his notebooks – admired fundamental papers Bolzano had published in 1816/1817.

Bolzano held the chair in the philosophy of religion at the University of Prague from 1805, but his liberal, anti-militarist views led to his suspension from his post by the Habsburg authorities in 1819. He was placed under house arrest, prohibited from publishing, and had little or no contact with the principal centres of mathematical research.

There is no real evidence that Bolzano's far-sighted ideas on the foundations of analysis were widely appreciated during his lifetime, although a well-known contemporary review journal in 1821 contained a favourable and fairly detailed review of his earlier papers (see [41]).

Cauchy's definitions did not distinguish clearly between continuity at a point and continuity at *all* points throughout some (closed) interval.[9] The relationship between these concepts, and what became the modern terminology, was clarified only a couple of decades later, in papers by *Peter Gustav Lejeune Dirichlet* (1805-1859) and, decisively, by *Karl Weierstrass* (1815-1897) in his Berlin lectures. *Eduard Heine* (1821-1881), who had studied under Dirichlet in Berlin, was a third significant contributor to the modern formulations.

Between them, Cauchy and Bolzano contributed three results (whose proofs can be found in *MM* and [27]) that were critically important for the subsequent programme to justify Calculus techniques without resort to 'infinitesimals'. Leibniz, who saw continuous functions as representing geometric curves in the plane, would probably have dismissed all three as 'intuitively obvious', but their rigorous deduction from the definition of continuity given above is a different matter.

Cauchy and Bolzano sought to prove Leibniz' intuitive assumptions without any recourse to geometric representations. This was an important step towards freeing the Calculus from reliance on geometric intuition. But their work also highlighted the sense in which the nature of the underlying *continuum* needed to be specified clearly.

We list the three key results, taking the real-valued function f to be continuous on a bounded interval I of real numbers:

The first says that if f takes distinct values α, β at points a and b in I then it also takes all intermediate values.

The second states that for any infinite sequence of real numbers contained in I, an infinite number of them (an infinite *subsequence*) must 'bunch up' near some point and converge to it.

---

[9]The difference lies in the freedom of choice allowed when choosing $\delta > 0$ for a given $\varepsilon > 0$. Continuity of f at the point a allows us to choose δ which, for that *specific* a, ensures that $|f(x - f(a)| < \varepsilon$ when $|x - a| < \delta$. Thus different δ can be chosen for different points a; in other words, δ may depend on a. A stronger demand defines *uniform continuity* over an interval I : here the *same* $\delta > 0$ is expected to suffice for *all* a in I. The ambiguity of Cauchy's definition may be what led him to claim that, on a closed interval, the sum of a convergent series of continuous functions will be continuous. In fact, this result requires either uniform convergence of the series or uniform continuity of the functions. (These concepts were first formulated in the 1840s.) However, Cauchy continued to insist on his claim, even though Abel had constructed a counterexample in 1826. The reasons for what is traditionally regarded as Cauchy's 'error' have been widely discussed in the literature, for example in Appendix 1 of [29]. A recent account of the history of this controversy can be found at *arXiv:1704.07723*.

The third asserts that if f is continuous at all points of a closed bounded interval $[a, b]$ contained in I then the collection of values f takes is bounded (above and below) and f attains those bounds at some points of $[a, b]$.

More formally, these claims are expressed as follows:

(i) The Intermediate Value Theorem (IVT)

If the real-valued function f is continuous on an interval containing a, b and $f(a) < 0 < f(b)$ then $f(c) = 0$ at some point c between a and b.

(ii) The Bolzano-Weierstrass Theorem

Every bounded real sequence has at least one convergent subsequence.

(iii) Extreme Value Theorem

A continuous function defined on a closed bounded interval $[a, b]$ of real numbers is bounded and attains its bounds on the interval $[a, b]$.

To illustrate how the IVT relies on properties of the underlying number system, we apply the modern definition of continuity at a point to functions defined only on the *rational* numbers in the interval $(0, 1)$. Then the function $f(x) = x^2$ will be continuous at each rational r in $(0, 1)$. This follows because

$$|f(x) - f(r)| = |x^2 - r^2| = |x - r|\,(x + r) < 2\,|x - r|,$$

since both x and r lie between 0 and 1. Hence, given (rational) $\varepsilon > 0$, we have

$$|f(x) - f(r)| < 2\,|x - r| < \varepsilon$$

provided that $|x - r| < \delta$, where $\delta = \frac{\varepsilon}{2}$.

On the other hand, $f(x)$ takes the rational values $f(\frac{1}{4}) = \frac{1}{16}$ and $f(\frac{3}{4}) = \frac{9}{16}$, but *not* the intermediate value $\frac{1}{2}$, as there is no rational number whose square is 2. So the continuous function $g(x) = f(x) - \frac{1}{2}$ is negative at $\frac{1}{4}$, positive at $\frac{3}{4}$, but is never 0 at any rational point. Hence the IVT will *not* hold if the rational numbers are the underlying number system.

The *Bolzano-Weierstrass* theorem provides yet another way of expressing the completeness of the continuum. It features in Bolzano's 1817 paper, effectively as a lemma in his derivation of the IVT, but expressed somewhat differently from the statement we have given. The theorem was reformulated and proved by Weierstrass more than 20 years after Bolzano, and became known as 'Weierstrass' Theorem' until the 1870s, when Bolzano's notebooks were rediscovered more than two decades after his death.

The Extreme Value Theorem states that if f is continuous on a closed bounded interval I then the image of I under f is also a closed bounded interval. For this, it is crucial that the interval I is closed: for example, the function $f(x) = \frac{1}{x}$ is continuous on the interval $(0, 1)$, but takes arbitrarily large values when x approaches 0. The function $g(x) = x$ is bounded on $(0, 1)$, but it does not have a maximum or minimum on the interval $(0, 1)-$

it approaches 0 and 1 arbitrarily closely, but never reaches either, as its values lie strictly between 0 and 1.

These simple examples illustrate that the three above theorems are meaningful and depend on properties of the underlying number system. The Extreme Value Theorem was also critical for Cauchy's quest to provide firmer foundations for the key results of the Calculus, as we outline below.

Bolzano was quite explicit in his critique of any proof in analysis which depended *'on a truth borrowed from geometry'*. Before he proved the Intermediate Value Theorem as a direct consequence of his definition of continuity he stated his views plainly:

It is an intolerable offence against correct method to derive truths of pure (or general) mathematics (i.e. arithmetic, algebra, analysis) from considerations which belong to a merely applied (or special) part, namely, geometry.... A strictly scientific proof, or the objective reason, of a truth which holds equally for all quantities, whether in space or not, cannot possibly lie in a truth which holds merely for quantities which are in space.

He dismissed Newton's approach to the Calculus in similar fashion:

No less objectionable is the proof which some have constructed from the concept of the continuity of a function with the inclusion of the concepts of time and motion.... No one will deny that the concepts of time and motion are just as foreign to general mathematics as the concept of space.

Nevertheless, and despite the greater precision of Cauchy and Bolzano's approach, an *arithmetical* definition of the underlying continuum remained absent. For example, Cauchy's proof of the IVT simply assumed that any bounded increasing sequence $(a_n)_{n \geq 1}$ of points on the number line must converge to a limit as n grows. This is another version of the *completeness property* of the real number system that the above results served to bring more clearly into focus.[10] Various approaches to this question will be examined more closely in **Chapter 7.**

4. Derivative and integral

The above theorems lead quite painlessly to the principal results needed for the Calculus, culminating in the proof of the Fundamental Theorem of the Calculus, which clarifies how differentiation and integration are linked as inverse operations, as Leibniz had claimed.

Cauchy's definition of the derivative of a function f at a point a in some open interval I makes essential use of his notion of limit:

---

[10]There are many equivalent versions of the completeness property of the real number system. In *MM* we consider five of these in more detail.

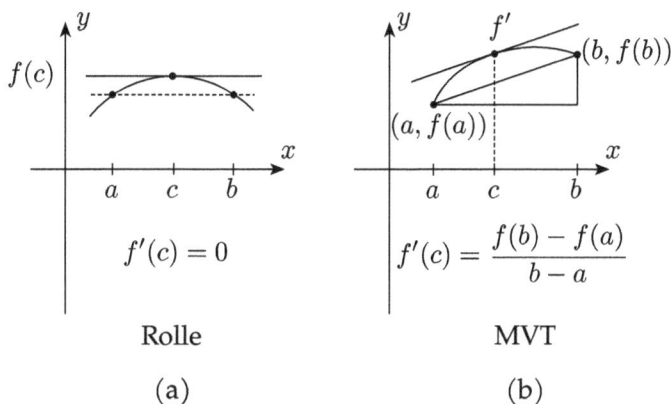

Figure 34. The Mean Value Theorem

If f is defined on $I = (a, b)$, the derivative of f at a, denoted by $f'(a)$, is given by $\lim_{x \to a} \frac{f(x) - f(a)}{x - a}$.

When this limit exists, f is said to be *differentiable* at a. Also, f is *differentiable on I* if it has a derivative at each point of I.

Note that the *slope function* for f, given by the ratio $\frac{f(x) - f(a)}{x - a}$ (the difference of the ordinates divided by the difference of the abscissae) makes sense only as long as $x \neq a$. This ratio is the slope of the *chord* to the graph of the function between the points $(a, f(a))$ and $(x, f(x))$. The slope function is not defined at a, but Cauchy *defines* the limit of these slopes as the slope of the *tangent* to f at a. In Newton's terminology it represents the *'ultimate ratio'* as x gets ever closer to a.

Cauchy's definition enabled him to prove the basic properties of derivatives – such as linearity and the product, quotient, inverse and chain rules – as simple consequences of the properties of limits.

Some properties of derivatives, however, have their roots in deeper properties of the number line. The first result reflecting this is again deceptively simple – it was stated in 1691 the Frenchman *Michel Rolle* (1652-1719), who was actually a vocal critic of both Newton and Leibniz' formulations of the Calculus. The second theorem (MTV) is an immediate consequence – again, proofs of these results can be found in *MM* and [27]. Figure 34 illustrates these theorems in turn.

Rolle's Theorem

If f is continuous on $[a, b]$ and differentiable on (a, b), and if $f(a) = f(b)$, then there is at least one point c in (a, b) such that $f'(c) = 0$.

The Mean Value Theorem

If f is continuous on $[a, b]$ and differentiable on (a, b), then there exists c in (a, b) with $f'c) = \frac{f(b)-f(a)}{b-a}$.

The Mean Value Theorem provides the key to many of the familiar applications of derivatives: finding the maximum or minimum values of a given differentiable function in some interval, finding its points of inflexion, applying rules (for example, the well-known *l'Hôpital's rule* for limits of quotients such as $\frac{\sin x}{x}$ at 0), and the term-by-term differentiation of power series that was used so extensively by Newton.

While Cauchy's definition of the derivative mirrors that of some of his predecessors, his definition of the integral was entirely novel. It brought into focus the seventeenth century geometric perceptions of the integral as the area under the graph of a curve, regarded as an infinite sum of infinitesimal slices, but without recourse to geometry or infinitesimals. As the latter had been criticised as imprecise and problematic, the preferred definition of the integral used by eighteenth century exponents of the Calculus was that integration simply constituted the 'inverse' of differentiation. This was expressed most directly by Cauchy's former teacher, *Lacroix*:

'the integral calculus is the inverse of the differential calculus, its object being to ascend from the differential coefficients to the function from which they are derived'.

This definition of the integral as an *antiderivative* – given a function f, find a function F that has derivative $F' = f$ (much as Newton had argued for 'finding the fluent from the fluxion') – held sway until Cauchy took on the task of defining 'the area under a curve' in a more rigorous *arithmetical* fashion.[11]

Cauchy restricted his attention to continuous functions defined on a closed interval $[a, b]$. He realised that not every 'area under the graph' of a function could be expressed as the difference of the values of an antiderivative at b and at a. So, instead of using the antiderivative as a *definition* for the integral, he sought to make it a *theorem* that the area under the graph of a curve could be found as this difference. This expression of the inverse relationship between derivative and integral has become known as the *Fundamental Theorem of Calculus*.

For a continuous function f defined on the closed interval $[a, b]$, Cauchy considered a partition of the interval $a = x_0 < x_1 < x_2 < ... < x_n = b$ and

---

[11]Note that antiderivatives are not uniquely defined: if two functions differ by a constant their derivatives are equal. For a practical illustration, recall that velocity was defined as the derivative of position (or 'distance travelled') by Newton. So two vehicles travelling in the same direction for the same time period (and with the same velocity curve) will only arrive at the same point if their starting positions are the same!

formed the sum (of 'areas of rectangles')

$$S_n = \sum_{k=0}^{n-1} f(x_k)(x_{k+1} - x_k)$$

He showed that the sequence $(S_n)_n$ will converge to a limit if the *mesh* $\delta = \max_{k<n}(x_{k+1} - x_k)$ of the partition converges to 0 when $n \to \infty$, *independently* of the partitions $(x_k)_k$. This proof required considerable effort, but the payoff was immediate: Cauchy could now use the limit as his *definition* of the integral, thus showing that the integral can be defined as the limit of sums that approximate the 'area under the curve'.

Cauchy's decision to turn the existence of the limit into his definition of the 'area under the graph', was a masterstroke. It meant that the existence of the integral over the interval $[a, b]$ was guaranteed for any function f continuous on $[a, b]$, removing any reliance on first finding an antiderivative – which, as he knew, could not be found via a closed formula in some cases, such as $f(x) = e^{-x^2}$.

The Extreme Value Theorem now sufficed (see *MM*) for Cauchy's proof of the main result of the Calculus, showing that integration and differentiation are 'inverses' of each other:

The Fundamental Theorem of the Calculus:

Suppose that f is continuous of $[a, b]$. For any x in $[a, b]$, the function F defined on $[a, b]$ by $F(x) = \int_a^x f(t)dt$ is an antiderivative of f: for x in (a, b), $F'(x) = f(x)$.

Throughout the nineteenth century, increasingly sophisticated versions of the integral were developed, particularly by *Bernhard Riemann* (1826-1866)—who had studied with Gauss and Dirichlet—but also later by *Camille Jordan* (1838-1922). This led to the development of numerous functions that illustrated with increasingly clarity why the 'number line' was in need of a rigorous arithmetical description. See *MM* for a summary account of these developments.

CHAPTER 7

Number Systems

If a man's wit be wandering, let him study the mathematics; for in demonstrations, if his wit be called away never so little, he must begin again.

Sir Francis Bacon, 'Of Studies' in: *Essays*, 1597

Summary

Chapters 1-4 described key aspects of the historical evolution of the number concept up to the early nineteenth century, largely driven by the realisation that the solution of various problems (first posed verbally but increasingly expressed via various types of equation) would lead far beyond the Pythagorean concept of 'multiples of the unit' as the only entities regarded as 'numbers'. By the early 1800s, these developments had led to the *Fundamental Theorem of Algebra*, clarifying the basic structure of polynomials. This had required the extension of the number concept to complex numbers.

Chapters 5 and **6** outlined two key episodes in the history of mathematicians' struggle with the concept of *infinity*, arising in the analysis of 'continuous magnitudes' to understand the dynamics of instantaneous change and the summation of infinitely many terms. The development of *Calculus* techniques addressed both problems with a considerable measure of success, but only by postulating *infinitesimal quantities* whose existence could not be established convincingly. In the nineteenth century, these quantities were finally 'banished' through conceptual advances made mainly in Paris and Berlin, giving a central role to the notions of *limit* and *continuity* of functions. This, in turn, brought into much sharper focus the question of the nature of the underlying *continuum*, represented by the geometric 'number line' inherited from geometry.

The present chapter deals with several aspects of *number systems*. We review the familiar systems of numbers encountered on this journey. The natural (or counting) numbers are taken for granted, but we explore their *structure* a little further, highlighting, in particular, the *Principle of Induction.* Our initial interest in induction is as a proof technique; in **Chapter 8** it will be seen as a fundamental property of the set of natural numbers. Next we outline the scheme devised by *Richard Dedekind* (1831-1916) to produce rigorous definitions of integers and rationals and their arithmetic, using only

https://doi.org/10.11647/OBP.0236.07

the familiar properties of the natural numbers, while showing that these properties are inherited and extended in the new number systems.

The next two sections contrast the differing approaches of Dedekind and *Georg Cantor* (1845-1918) to the development of the *real number system*. Both started with the rationals. Dedekind emphasised the analogy with the order properties of the line, while Cantor employed (classes of) Cauchy sequences to define each real number, simplifying the extension of arithmetical properties from the rationals to the reals.

Finally, we consider infinite decimal expansions, a concept that, while familiar from 'recurring' decimal expansions encountered at school, has hidden depths that repay closer study on several counts. Rationals and irrationals give rise to two distinct (infinite) classes of expansions, and we indicate why the irrationals seem to be 'more numerous'. We also identify *constructible, algebraic* and *transcendental* numbers, identifying how solutions of the 'three famous problems' of antiquity fit into these classes.

1. Sets of numbers

We have so far attached names to six different types of number, developed and explored over two millennia—albeit with rather variable degrees of precision. I will employ the *language of sets* to describe the collection of numbers in each category. This practice is of relatively recent origin, but is well-established in common parlance today—not least because of its (initially quite controversial) introduction into primary school teaching in many developed countries during the 1960s. I will use this 'naive set theory' informally, primarily to have a convenient notation and terminology, rather than engage with the logical formalities of any abstract *theory of sets*.

The six collections of numbers highlighted below are examples of *sets* whose members are various types of number. But we can imagine many different collections of objects, concrete or abstract, and it is convenient to have a simple terminology to identify them.[1]

Here is a brief review of this terminology and notation:

By a set S we will understand any collection of distinct objects (mental or physical), together with a *membership rule* enabling us to decide whether a given object x is a member of the collection or not. If it is, we write $x \in S$ (expressed variously as 'x is an element (or member) of S', or 'x is in S', or 'x belongs to S'); if it is not, we write $x \notin S$ ('x is not in S'). The set S is an object in its own right —so S can itself be a member of some other set.

---

[1] In 1895, *Georg Cantor* defined the term *set* as follows: 'By a set M we understand every gathering together into a whole of definite, distinct objects m (which are called the 'elements' of the set) of our perception or of our thought.'

For a finite set S one might simply list its elements—if these are a, b, c, write $S = \{a, b, c\}$. We write $\{x\}$ for the set containing just the element x. Such a set is called a *singleton*.

We define the *empty set* $\varnothing$ as a set that has *no* elements—this can be done by using a contradictory membership rule: for example, if $P(x)$ is a statement involving the variable x, then we may write $\{x : P(x)\}$ to specify the set of all possible elements x for which the statement $P(x)$ is true. Then $\varnothing = \{x : x \neq x\}$, specifies 'the set of all x such that x does not equal x'. Since there are no such x, the set $\varnothing$ has no elements. A set is *non-empty* if it is not the empty set.

Comparison of sets introduces further basic notation and terminology:

Given two sets A, B, they are *equal* (written $A = B$) if they contain exactly the same elements; that is, every element of A is an element of B and vice versa.

Call A a *subset* of B (and write $A \subseteq B$) if every element of A is an element of B. This includes the possibility that the two sets are equal. The empty set $\varnothing$ is a subset of every set.

Call A a *proper subset* of B (written $A \subset B$) if every element of A is an element of B, but not vice versa – so that B contains elements that are not in A. In this case we write the *set difference* as $B \backslash A$. For example, $A = \{1, 2\}$ is a proper subset of $B = \{1, 2, 3\}$, and $B \backslash A = \{3\}$.

We write $A \cup B$ for the *union* of sets A and B. This is the set whose elements belong to either A or B (or both). The set of all elements that A and B have in common is called their *intersection*, written as $A \cap B$. If two sets A and B have no elements in common (so that $A \cap B = \varnothing$) they are *disjoint*.

We also extend these definitions to a *sequence* $(A_n)_{n \geq 1}$ of sets, denoting the union by $\cup_{n \geq 1} A_n$ and the intersection by $\cap_{n \geq 1} A_n$.

The types of number encountered so far are:

Positive whole numbers—evolving from the 'counting numbers', or 'the unit and its multiples', and now known more formally as the *natural numbers*. The set of all natural numbers is denoted by $\mathbb{N}$.

Integers—extending the natural numbers to include zero and negative numbers. The resulting set of all integers is written as $\mathbb{Z}$.

Rational numbers—deriving from the Greek notion of 'commensurable lengths', these are expressed as ratios of integers without common factors; or more colloquially as 'fractions in lowest form'. We write the set of all rational numbers as $\mathbb{Q}$.

Irrational numbers—deriving from 'incommensurable lengths' to distinguish them from the rationals. We have yet to give them a satisfactory definition as numbers.

Real numbers—described, until now, only by reference to points that can be marked on an unlimited geometric 'number line'. We have claimed that this comprises all rational or irrational numbers, taken together. The set of all real numbers will be denoted by $\mathbb{R}$.

Complex numbers—depicted, by analogy, as points in the plane, they may be regarded more formally as *ordered pairs* of real numbers, for which *sums* and *products* are formed by Hamilton's explicit rules. Their full definition will therefore follow without difficulty once the concept of 'real number' has been defined consistently and independently of any geometric description. The set of all complex numbers is denoted by $\mathbb{C}$.

The sets $\mathbb{N}, \mathbb{Z}, \mathbb{Q}, \mathbb{R}$ are today firmly embedded in our culture and daily experience. While the set $\mathbb{C}$ may be less familiar, complex numbers have been central to many areas of mathematics and to progress in several areas of science and engineering, especially in applications of electromagnetism and in electronics, for at least the past 150 years. We will now look at these five sets more abstractly, as objects in their own right, and ask how they relate to each other.

Various examples have illustrated how the basic *arithmetical operations* $(+, \times)$ and their inverses $(-, \div)$ are applied in computations and to obtain the solutions of various types of equations. We also know how the *order relation* $(<)$ is applied to compare natural numbers, integers, and rational or real numbers, and how this ordering interacts with addition and multiplication. It will be shown (in the Appendix to **Chapter 8**) that the system of complex numbers as a whole *cannot* be ordered in a way that is compatible with addition and multiplication.

As far as the *elements* of the sets are concerned, we can 'list' the set of integers as a doubly infinite sequence $\mathbb{Z} = \{..., -3, -2, -1, 0, 1, 2, 3, ...\}$, while the set $\mathbb{Q}$ of rationals can, in turn, be regarded as the collection of all fractions $\frac{m}{n}$, where m is any integer and n a natural number having no factors in common with $|m|$, so that the fraction is expressed in its 'lowest form'.

Observe that sums and products of natural numbers are again natural numbers (so that performing addition and multiplication in $\mathbb{N}$ cannot take us beyond $\mathbb{N}$). The same is true for $\mathbb{Z}$ and $\mathbb{Q}$: for example, the sum of two rational numbers, expressed as ratios (in lowest form) of integers, $\frac{m}{n}$ and $\frac{p}{q}$, is $\frac{mq+np}{nq}$, which is again a ratio of two integers, as is their product $\frac{mp}{nq}$. We say that both number systems are *closed* under addition and multiplication. Unlike $\mathbb{N}$, the set of integers $\mathbb{Z}$ is also closed under *subtraction*, but not under *division* (these are the *inverse* operations to addition and multiplication), while $\mathbb{Q}$ is closed under subtraction, and $\mathbb{Q}\backslash\{0\}$ is closed under division. All this (though perhaps not the terminology) should be entirely familiar from schooldays.

However, the product of two *irrational* numbers need not be irrational – consider $(\sqrt{2})(\sqrt{2}) = 2$, for example. Since it is not closed under multiplication, the set of all irrational numbers will *not* be treated as a *number system* in the same fashion as $\mathbb{N}$, $\mathbb{Z}$, or $\mathbb{Q}$. A major task in this Chapter will be to incorporate this set, together with $\mathbb{Q}$, in the single number system $\mathbb{R}$, whose elements are described *arithmetically*. Two solutions to this occupy Sections 4 and 5, while in the final three short sections we briefly consider other methods of distinguishing between rationals and irrationals as well as between different types of irrational numbers.

2. Natural numbers

2.1. The principle of induction. We have treated the set of natural numbers $\mathbb{N} = \{1, 2, 3, ...\}$ as given: in other words, it simply exists and we can identify its elements successfully. This assumption is implicit in the process of counting, and corresponds to a perception of $\mathbb{N}$ that was universal from the time of the Pythagoreans until well into the nineteenth century. Until then, mathematicians saw no need to devise a more formal framework in which to derive elementary properties of the set $\mathbb{N}$. This point of view was perhaps expressed most bluntly in a famous phrase attributed to the influential *Leopold Kronecker* (1823-1891) in Berlin:

Die ganzen Zahlen hat der liebe Gott gemacht, alles andere ist Menschenwerk. (God created the natural numbers, all else is the work of man).

However, $\mathbb{N}$ is an *infinite* set, as every natural number n has an immediate successor $n + 1$, as was observed in the **Prologue**. Therefore specific techniques are needed to analyse it (and its arithmetic) more fully to answer certain types of question. The most important of these is the *Principle of Induction*. Intuitively, this seeks to make the statement 'and so on' more precise. It can be pictured as consecutive dominoes falling in a never-ending row, or as climbing an infinite ladder step by step, from the bottom rung. Or simply by seeking to count numbers one-by-one 'forever'.

As a method of proof this principle has a long history, but it was typically used *implicitly*, and often inconsistently.[2] Perhaps the most explicit early example of its use can be found in the work of *Levi ben Gershon* (1288-1344) – see [**25**] for details. In the seveteenth century Blaise Pascal used induction explicitly when justifying the properties of his well-known Pascal

---

[2]Mathematical induction concerns a specific method of proof, and should not be confused with *inductive logic*, which, in philosophy, is concerned with measures of evidential support for assertions (see the entry on *Inductive Logic* in the online *Stanford Dictionary of Philosophy*).

triangle (see Figure 30, page 122). Fermat frequently used the closely related method of 'infinite descent', using the fact that for any n there are only finitely many different natural numbers less than n. However, it was only at the end of the nineteenth century that the Induction Principle was taken as an *axiom* in the work of *Giuseppe Peano* (1858-1932). This will be discussed in **Chapter 8**.

The basic idea is simple: suppose that we can verify that a given statement $P(n)$ about the natural number n holds when $n = 1$ (i.e. $P(1)$ is true) and that we can prove the following *inductive step*: for any n, the truth of $P(n)$ implies the truth of $P(n+1)$. Then the Principle claims that $P(n)$ must hold for *all* natural numbers.

The following two simple examples illustrate the use of the Induction Principle:

(i) $1 + 3 + 5 + ... + (2n - 1) = n^2$ for every n in $\mathbb{N}$. (Recall the 'pebble proof' of this claim in **Chapter 1**)

Proof: Let $P(n)$ be the statement: $1+3+5+...+(2n-1) = n^2$. Then $P(1)$ becomes the statement $1 = 1$, which is true.

Next, take n in $\mathbb{N}$ and assume that $P(n)$ holds. This implies $P(n+1)$, because

$$1+3+...+(2n-1)+(2n+1) = n^2 + (2n+1) = (n+1)^2 = P(n+1).$$

By the Induction Principle, $P(n)$ holds for every natural number.

(ii) *Bernoulli's inequality:* Fix a real number $a > -1$. Then $(1+a)^k \geq 1+ka$ for every natural number k.

Proof: For each $k \geq 1$, let $P(k)$ be the statement: if $a > -1$, then $(1 + a)^k \geq 1 + ka$.

Again, $P(1)$ is true: $(1 + a)^1 = 1 + a$.

Assume that $P(k)$ holds. Note that $a > -1$ ensures that $(1 + a)$ is positive. As $a^2 \geq 0$, we obtain

$$(1+a)^{k+1} = (1+a)^k(1+a) \geq (1+ka)(1+a)$$
$$= 1 + (k+1)a + ka^2$$
$$\geq 1 + (k+1)a,$$

so that $P(k + 1)$ holds.

Again, by induction, $P(n)$ holds for every natural number n.

Bernoulli's inequality enables us to verify formally that $\lim_{n\to\infty} x^n = 0$ when $|x| < 1$. (See Achilles and the Tortoise in **Chapter 5**.)

Proof: As $|x| < 1$, we can write $|x| = \frac{1}{1+b}$ for some $b > 0$. Let $\varepsilon > 0$ be given. By Bernoulli's inequality,

$$|x|^n = \frac{1}{(1+b)^n} \leq \frac{1}{1+nb}.$$

Hence $|x^n| < \frac{1}{nb} < \varepsilon$ for all $n \geq N$, provided that we choose $N > \frac{1}{b\varepsilon}$.

So $\lim_{n \to \infty} x^n = 0$.

The Induction Principle implies another property of $\mathbb{N}$ that may seem blindingly obvious:

Well-Ordering Property (WO) of $\mathbb{N}$

Every non-empty subset of $\mathbb{N}$ has a least element.

Proof: We prove this by contradiction, using the Induction Principle. Suppose that (WO) is false for the set $\mathbb{N}$. Then it would have a non-empty subset S with no least element. Let $P(n)$ be the statement "$i \notin S$ for all $i \leq n$". In other words, $P(n)$ holds if S does not contain any of the natural numbers $1, 2, 3, ..., n$. In particular, $1 \notin S$, since if it were in S it would be its least element. Therefore $P(1)$ is true. For the inductive step, assume that $P(n)$ is true, so that none of $1, 2, 3, ..., n$ belong to S. In that case, $n+1$ cannot belong to S, since otherwise it would be its least element. But this means that $P(n+1)$ is true, and by induction, $P(n)$ holds for all n in $\mathbb{N}$. This means that S must be empty, contradicting our assumption that the (WO) is false. Therefore (WO) must hold for $\mathbb{N}$.

We have proved that the Induction Principle implies (WO). In fact the two are logically equivalent—we omit the other half of the proof.

In some contexts another version of induction is useful. The *Strong Induction Principle* uses the following induction step instead: assume that $P(m)$ is true for all $m \leq n$ and prove that $P(m+1)$ is true. If this can be shown, then the Strong Induction Principle asserts that $P(n)$ holds for all n in $\mathbb{N}$.

We will use this version of the Principle repeatedly below. The prefix 'strong' does not mean that we can prove more with this version than with that stated earlier. It refers to the fact that we make a *stronger assumption* in the induction step. The two versions of the Principle are logically equivalent (this is proved in *MM*). Note, however, that we have made no attempt to *prove* either version from earlier statements.

2.2. Prime numbers as building blocks. The set $\mathbb{N} = \{1, 2, 3, ...\}$ is familiar from primary school. We learn early on that amongst the natural numbers there are some special ones: *prime numbers*. These are the natural numbers *greater than* 1 that have no proper divisors. In other words, a prime number p is divisible only by 1 and by itself. Dividing p by 1 changes nothing, and dividing p by itself results in 1.

All other natural numbers $n > 1$ are called *composite*. So a composite number n has a proper divisor a, that is, a natural number greater than 1 but less than n (written $1 < a < n$), such that $n = a \times b$ for some natural number $b > 1$. The divisors a and b are called *factors* of n.

For a long time 1 was treated as a prime number (since it certainly has no factors!) and this practice fits well with the use of the term in ordinary language. But this ended more than a century ago. A mathematical definition should be judged by its usefulness in identifying the nature of the objects being discussed. What we call the *Fundamental Theorem of Arithmetic* below will highlight one rather important reason why it is better *not* to include 1 among the prime numbers. So we will insist that a prime number must be greater than 1.

The first significant general result learnt at school about prime numbers is that any natural number greater than 1 has a *prime divisor*. If n is prime, it is its own prime divisor. If n is composite, then it must have at least one prime factor.

To see why this must be true we will use strong induction: suppose we have proved that every natural number less than some number n has a prime divisor. If n is prime there is nothing to prove. If n is composite then $n = a \times b$, where both factors a and b are greater than 1. But this means that $a < n$, so that a has a prime divisor by our inductive assumption. Call this divisor p. But if p divides a, it also divides $a \times b = n$. This means that n has a prime factor. Hence by strong induction every natural number, has at least one prime divisor.

This result illustrates how to decompose any natural number n into prime factors. If n is prime, nothing needs to be done. If n is composite it has at least one prime factor, as shown above. Suppose now that p_1 is the smallest of these, so that we can write $n = p_1 \times n_1$ for some natural number n_1 strictly between 1 and n. If n_1 turns out to be prime, we are done. Otherwise, it has a smallest prime factor p_2 and $n_1 = p_2 \times n_2$ for some natural number n_2 strictly between 1 and n_1. Continuing in this way, we obtain a strictly decreasing sequence $n_1, n_2, \ldots$ of numbers between 1 and n and, since n is finite, this must be a *finite* sequence (Fermat's method again): after finitely many steps (k say) we will have $n_{k-1} = p_k \times n_k$ with p_k prime and $n_k = 1$. It follows that $n = p_1 \times p_2 \times .. \times p_k$.

This simple procedure provides a *prime factorisation* of n.

Prime numbers are the basic building blocks of $\mathbb{N}$. The usefulness of this statement turns on the important question whether the above decomposition of n into prime factors is *unique*. If we were to treat 1 as a prime, then there are obviously many prime factorisations of each natural number n, since the factor 1 can be included as many times as we like. Thus, to have any hope of obtaining uniqueness, 1 must be excluded from the primes.

The factors can be shuffled without changing the product. This follows because multiplication in $\mathbb{N}$ is *commutative*: for any m, n in $\mathbb{N}$ we have $n \times m = m \times n$.

Also, some of the factors might 'repeat', i.e. occur several times. To avoid such trivial variations in our representation we write the product of the prime factors of n as follows from now on:

$$n = p_1^{k_1} p_2^{k_2} ... p_j^{k_j}.$$

Here the prime factors are given in increasing order $p_1 < p_2 < ... < p_j$ with k_l being the number of times that p_l occurs in the product, for each $l = 1, 2, ..., j$. For example, $584 = 2^3 \times 73$, while $2520 = 2^3 \times 3^2 \times 5 \times 7$.

The claim, to be proved below, is that any natural number n has a unique prime factorisation in this form.

First, however, we explore prime numbers a little further.

We might ask *how many* prime numbers there are. Here is the list of prime numbers below 100:

$2, 3, 5, 7, 11, 13, 17, 19, 23, 29, 31, 37, 41, 43, 47, 53, 59, 61, 67, 71, 73, 79.83, 89, 97.$

These 25 numbers 'thin out' somewhat as the numbers grow: there are four primes below 10, but only two between 80 and 90, and one after that. Since it becomes 'more difficult' for a number to be prime as the number of possible divisors increases, it might be tempting to guess that the list of all primes could stop somewhere. But Euclid showed that this guess would be wrong.

We may paraphrase the statement of Euclid's *Elements*, Book IX, Proposition 20) as follows:

The number of primes is not finite.

The proof is very simple. Take any finite collection of primes,

$$\{p_1, p_2, p_3, ..., p_k\}$$

and form the number $n = (p_1 \times p_2 \times p_3 \times ... \times p_k) + 1$. If n is prime, it is a new prime greater than all the p_i. If n is composite, it has a prime factor, p say, as was shown above. This p cannot be one of the p_i ($i = 1, 2, .., , k$): if it were, it would divide both n and the product $(p_1 \times p_2 \times p_3 \times ... \times p_k)$ and so p would divide their difference $n - (p_1 \times p_2 \times p_3 \times ... \times p_k) = 1$, which is impossible. This shows that p is a prime not in the above list, and therefore no finite collection of primes can exhaust the collection of all primes.

A remarkable, and much stronger, result was published in 1837 by Lejeune Dirichlet:

If a, d are natural numbers with no common factors, there are infinitely many primes in the infinite arithmetic progression

$$a, a + d, a + 2d, a + 3d, ..., a + nd, ...$$

See *MM* for examples and some consequences of this theorem. It represents most of what we know about infinite collections of primes that follow a given pattern—the search for a 'formula' that would provide the value of the nth prime has long been abandoned as hopeless!

A less ambitious question is how the primes are *distributed* among the elements of $\mathbb{N}$. Since we cannot expect a formula for the number $\pi(N)$ of primes $p \leq N$, mathematicians have sought, instead, to establish *asymptotic estimates* for $\pi(N)$. These are estimates for $\pi(N)$ when N grows very large. They included a correct conjecture made in the early 1790s by the teenage Gauss[a], who had examined the list of primes below 3 million!

It was not until 1896 that Gauss' conjecture was verified (independently) by two mathematicians: the French *Jacques Hadamard* (1865-1963) and the Belgian *Charles de la Vallee Poussin*[b] (1866-1962). It remains one of the highlights of nineteenth century mathematics:

The Prime Number Theorem

For large N, $\pi(N)$ behaves asymptotically like $\frac{N}{\log N}$. More precisely,

$$\lim_{N \to \infty} \frac{\pi(N)}{\left(\frac{N}{\log N}\right)} = 1.$$

Loosely speaking, this says that, for large enough N, the proportion $\frac{\pi(N)}{N}$ of primes in $\{1, 2, 3, ..., N\}$ is 'close' to $\frac{1}{\log N}$. Here, as in **Chapter** 5, log denotes the natural logarithm.

---

[a]Near the end of his life Gauss stated that he had reached his conjecture in 1792 or 1793 (aged 15 or 16). But he did not publish his findings during his lifetime. His motto was *Pauca sed matura* (Few, but ripe) and he was unwilling to claim results for which he did not have a full proof. The first (slightly different) published conjecture of an asymptotic estimate for $\pi(N)$ was given by the French mathematician *Adrien-Marie Legendre* in 1798.

[b]He was ennobled by the King of Belgium for his feat, becoming *Charles-Jean Étienne Gustave Nicolas, baron de la Vallée Poussin.*

2.3. Uniqueness of prime factorisation. We have shown that every natural number n can be represented as a product of its prime factors in the form

$$n = p_1^{k_1} p_2^{k_2} ... p_j^{k_j}$$

We now want to show that this representation is unique (up to shuffling of the factors). For this we prepare the ground a little, and show how a very familiar concept plays a crucial role. The *greatest common divisor* (gcd) of two natural numbers a, b is the largest natural number d that divides both a and b.

To show that the gcd d of two natural numbers a, b always exists we characterise it slightly differently.

Define d' as the smallest natural number such that there are integers s, t (positive or negative) such that
$$d' = sa + tb.$$
For any pair of natural numbers a, b there is always at least one pair of integers s, t making $sa + tb$ a natural number. Let L be the set of all natural numbers $sa + tb$ found in this way. Therefore, d' exists by the Well-Ordering Principle, since it is the least member of the non-empty subset L of $\mathbb{N}$.

Any common divisor of a and b clearly also divides d'. We now show that d' is itself a common divisor of a and b, which ensures that d' coincides with the greatest common divisor d of a and b.

Using 'long division', write $a = qd' + r$ for some integer q (the 'quotient') and an integer r with $0 \leq r < d'$ (the 'remainder'). Therefore
$$r = a - qd' = a - q(sa + tb) = (1 - qs)a + (-qt)b = ua + vb,$$
where we have $u = 1 - qs$ and $v = -qt$. Here u and v are integers, so r has been written in the form $ua + vb$ for some integers u, v. By definition, d' is the *smallest* natural number that can be written in this form. On the other hand, by our construction, $0 \leq r < d'$. This means that $r = 0$. Hence $a = qd'$, so d' divides a. A similar argument holds if we reverse the roles of a and b. Thus $d' = d$ is the gcd of a and b. This confirms that the gcd d of two natural numbers a and b always exists and can be written as $d = sa + tb$ for some integers s, t.

It follows, in particular, that if a and b are *relatively prime* (a and b have no common divisors other than 1), then we can always find integers s, t such that

$$sa + tb = 1.$$

These simple facts are used to prove a key result due to Euclid. First observe that if a natural number n is a divisor of the product ab of two natural numbers, then n need not divide either a or b. For example: 12 divides $10 \times 6 = 60$, but it divides neither 10 nor 6. Euclid shows that this *cannot* happen for a prime.

Euclid's Lemma:

If p is prime and divides the product ab, then p divides at least one of a and b.

Proof: Suppose that p is a prime divisor of the product ab. If p does *not* divide a, it has no factors in common with a (since p is prime). Therefore the gcd of a and p is 1, which means that we can find integers s, t satisfying $sa + tp = 1$.

Multiply this equation by b to obtain $sab + tpb = b$. But p divides ab, so $ab = cp$ for some natural number c. Hence $b = (sc + tb)p$. We have shown that p is a divisor of b and this proves Euclid's lemma.

Using induction, Euclid's lemma can be extended to products of many factors:

Corollary:

If p divides the product $a_1 a_2 ... a_k$, then p must divide at least one of the factors a_i for $i = 1, 2, ..., k$.

The simple proof is left to the reader. This completes the preliminaries.

Uniqueness of prime factorisation:

The prime decomposition $n = p_1^{a_1} p_2^{a_2} ... p_k^{a_k}$ of the natural number n is unique.

Proof: Suppose that there are two such decompositions:

$$n = p_1^{a_1} p_2^{a_2} ... p_k^{a_k} = q_1^{b_1} q_2^{b_2} ... q_l^{b_l}$$

where the p_i, q_j are primes, arranged in increasing order, for $i = 1, 2, ... k$ and $j = 1, 2, ... l$ respectively. Then each q_j divides the product $p_1^{a_1} p_2^{a_2} ... p_k^{a_k}$, hence it divides one of the p_i, where $i \leq k$, and each p_i divides the product $q_1^{b_1} q_2^{b_2} ... q_l^{b_l}$, hence divides one of the q_j. Since all are prime, it follows that each p is one of the q and vice versa. So $l = k$ and $p_j = q_j$ for each $j \leq k$, since the two products are in the given form.

It remains to show that the powers also correspond: if for some $j \leq k$ we have $a_j > b_j$, divide both expressions for n by $p_j^{b_j}$ and compare the resulting products

$$p_1^{a_1} p_2^{a_2} ... p_{j-1}^{a_{j-1}} p_j^{a_j - b_j} p_{j+1}^{a_{j+1}} ... p_k^{a_k}$$

and

$$p_1^{b_1} p_2^{b_2} ... p_{j-1}^{b_{j-1}} p_{j+1}^{b_{j+1}} ... p_k^{b_k}.$$

The two products are equal by assumption, but the first is divisible by p_j (since $a_j > b_j$) while the second product is not divisible by p_j. The contradiction shows that $a_j > b_j$ is impossible, and reversing the roles of the two expressions shows that $b_j > a_j$ is also impossible. So each $a_j = b_j$ and the two prime factorisations of n are identical, as claimed.

The uniqueness of prime factorisation is also known as the *Fundamental Theorem of Arithmetic,* since it gives us a complete description of the structure of the set $\mathbb{N}$ of natural numbers: each natural number can be obtained uniquely as a product of powers of primes, arranged in increasing order.

3. Integers and rationals

A trivial extension of set $\mathbb{N}$ of natural numbers results from adding the number 0 to it. We denote the result by $\mathbb{N}_0 = \mathbb{N} \cup \{0\}$; alternatively, we list its

Figure 35. Richard Dedekind, by an unknown photora-pher, 1900s[3]

elements as the sequence $\{0, 1, 2, 3, ..., n, ...\}$ where the difference between successive entries is always 1. Historically, 0 appeared much later than the 'counting numbers'; but it feels more logical to include it at the outset if we wish to investigate the arithmetical structure of $\mathbb{N}$: 0 is the *neutral* element for addition, since $n + 0 = n$ for any natural number n. In the same way, 1 is the neutral element for multiplication ($n \times 1 = n$ for all natural numbers n). We will take for granted the number system $(\mathbb{N}_0, +, \times)$; that is, the set $\mathbb{N}_0$ together with the arithmetical operations of adddition and multiplication. These are assumed to satisfy the following 'laws of arithmetic':

(i) *commutative:* $n + m = m + n, nm = mn),$

(ii) *associative:* $(m + n) + p = m + (n + p),$

(iii) *distributive:* $m(n + p) = mn + mp.$

Moreover, the set $\mathbb{N}_0$ is *ordered:* write $n < m$ if the equation $n + k = m$ has a solution for some $k \neq 0$ in $\mathbb{N}_0$. Also write $n \leq m$ if either $n < m$ or $n = m$. We could picture this on the 'geometric' number line by saying that either m and n coincide or m 'lies to the right' of n.

In 1854 this represented the starting point for Richard Dedekind in his quest to show rigorously how the familiar number systems we have described can be constructed directly from $(\mathbb{N}_0, +, \times, <)$, in terms both of definition of their elements and extension of the arithmetical operations and the order relation. Dedekind viewed mathematical progress as arising from

---

[3]https://commons.wikimedia.org/wiki/File:Richard_Dedekind_1900s.jpg

'free creations of the human spirit and the constraints imposed by logical necessity', which, for him, meant that new mathematical objects should *'always arise in a natural way from the current state of mathematical knowledge'* (the quotations are taken from [6]). In particular, extending numbers systems from $\mathbb{N}_0$ would, in the first place, require the *'unlimited completion'* of the arithmetical operations $(+, -, \times, \div)$ plus powers and roots.

An exhaustive treatment of the extensions may be found in the classic text [30], published in 1930 by *Edmund Landau* (1877-1935), where the required constructions are undertaken with the utmost rigour. In fact, Landau goes further: he shows how—as we will consider in the next chapter—the natural numbers themselves can be defined abstractly, how each extension to the next larger class (integers, rationals, real numbers and complex numbers) proceeds, how the arithmetical operations and order relations can be extended consistently in each case and how everything fits together.

Landau's Preface makes clear that he regards the contents of his book as wholly elementary, and presents it as essential background training for any mathematics student, to be read carefully at least once—after which the details can be forgotten!

My book is written, as befits such easy material, in merciless telegram style ("Axiom," "Definition," "Theorem," "Proof," occasionally "Preliminary Remark")... I hope I have written this book in such a way that a normal student can read it in two days. And then (since he already knows the formal rules from school) he may forget its contents.

Owing to Landau's highly 'telegraphic' and rigorous style, the book was received by his intended readership with rather less enthusiasm than he might have expected—perhaps for the reasons he mentions! Rather than repeat the exercise, I will largely omit detailed proofs in what follows below.

3.1. Dedekind on integers. To arrive at a number system that allows unlimited application of the four arithmetical operations $(+, -, \times, \div)$ the system $(\mathbb{N}_0, +, \times)$ must be extended twice: first, to allow us to *subtract* numbers without restriction, and then, from this extension, to allow *division* by any non-zero number. Dedekind's approach to both extensions was essentially the same: the process of finding the *difference* or the *quotient* of two natural numbers can be represented by considering *pairs* of numbers, i.e. elements of the *Cartesian product*, much as Hamilton had done in extending addition and multiplication from $\mathbb{R}$ to $\mathbb{C}$.[4]

---

[4]The *Cartesian product* of any two sets A and B is defined as the set of all *ordered pairs* (a, b) of elements $a \in A$ and $b \in B$. We write this as $A \times B = \{(a, b) : a \in A, b \in B\}$. A non-empty subset of the Cartesian product $A \times B$ is called a *relation*, since it describes ways of associating elements of the two sets with each other.

In each case a novel aspect of Dedekind's definitions was the explicit use of what is now called an *equivalence relation*.[5] This enabled him to provide unambiguous definitions of the 'new' number concepts he wished to display, using only the familiar properties of natural numbers. While the idea underlying equivalence relations can be found in specific contexts well before Dedekind, the definitions given below— never published by him— probably constitute the first consistent use of the method in general (see [13], p. 371), helping to usher in what *Bertrand Russell* (1872-1970) would later call '*definition by abstraction*'.

To define the elements of the first extension of $\mathbb{N}_0$ (to be denoted by $\mathbb{Z}$, whose elements he called *integers*) Dedekind used ordered pairs of elements of $\mathbb{N}_0$. He defined the operations of *addition* and *multiplication* for members of the set S of ordered pairs as follows:

$$(m, n) + (p, q) = (m + p, n + q), \quad (m, n) \times (p, q) = (mp + nq, mq + np).$$

We use the same symbols $+, \times$ on both sides, but the pairs on the right *define* the sum and product on the left.

The choice of definitions may be puzzling at first glance. But remember that $\mathbb{N}_0$ is being extended because it is not closed under *subtraction*. The pair (m, n) is intended to help define the *difference* $m - n$. For example, the above sum reflects the 'sum of differences' $(m + p) - (n + q) = (m - n) + (p - q)$.

Since we are working with elements of $\mathbb{N}_0$ we can use its arithmetic and order properties. If $n \leq m$ (so m 'lies to the right of' or equals n) the difference $(m - n)$ of the pair (m, n) will be in $\mathbb{N}_0$, so the pair will represent a familiar object. But if $m < n$ the pair will represent a *new* object. We might wish to define (e.g.) the pair $(2, 5)$ as representing -3, but this symbol will only be meaningful if we can extend the arithmetic of $\mathbb{N}_0$ to deal with such pairs.

Now note that the sum of the pairs $(5, 2)$ and $(2, 5)$ is the pair $(7, 7)$, whose 'difference' is 0. This is so for any pairs (m, n) and (n, m). The 'difference' is not uniquely defined in this way: taking $m = 5, n = 2$ produces this result, but so does $m = 4, n = 1$, since $(4, 1) + (1, 4) = (5, 5)$, for example. There is ambiguity in these definitions.

Dedekind resolved this by defining an *equivalence relation* on the set $S = \mathbb{N}_0 \times \mathbb{N}_0$. This would treat pairs with the same 'difference' as interchangable:

---

[5]Formally: an *equivalence relation* R on any set S is a subset R of the Cartesian product $S \times S$. If $(a, b) \in R$ we write $a \sim b$ ('a is related to b'). An equivalence relation must be:

(i) *reflexive*: $a \sim a$,

(ii) *symmetric*: if $a \sim b$ then $b \sim a$,

(iii) *transitive*: if $a \sim b$ and $b \sim c$ then $a \sim c$.

Clearly the equality relation $=$ has these properties. So an equivalence relation 'generalises' the concept of equality by *partitioning* the set S into *equivalence classes*. Members of the same class (*representatives* of the class) are treated interchangeably, and no two classes have any members in common (they are *disjoint*).

pairs (m,n) and (p,q) are related if $m + q = n + p$. We write this as $(m,n) \sim (p,q)$.

To check that this defines an equivalence relation on S, consider any pairs $r = (m,n)$, $s = (p,q)$, $t = (u,v)$ in S. Using the commutative and associative laws for addition, and cancellation in $\mathbb{N}_0$, we have:

(i) $r \sim r$ since $m + n = n + m$,

(ii) $r \sim s$ implies that $s \sim r$, since $m + q = n + p$ implies $p + n = q + m$, and

(iii) if for pairs r, s, t we have $r \sim s$ and $s \sim t$ then also $r \sim t$; for this note that $m + q = n + p$ and $p + v = q + u$. But then also $(m + q) + (p + v) = (n + p) + (q + u)$. So $m + v = n + u$, hence $r \sim t$.

Dedekind now defined the set $\mathbb{Z}$ of *integers* as the collection of *all equivalence classes* of pairs (where related pairs belong to the same class) under the relation $\sim$. Fortunately, it turns out that each class has exactly one representative which is either of the form $(n,0)$ or of the form $(0,n)$. The positive integer n is then the class containing $(n,0)$. If the pairs $(n,0)$ and (p,q) are equivalent, then $n + q = 0 + p$, so that the pair (q,p) is equivalent to $(0,n)$, since $q + n = p + 0$, and by the commutative law in $\mathbb{N}_0$ these two identities are the same. So we can define $-n$ as the class containing $(0,n)$. In this fashion we justify writing $\mathbb{Z}$ as the sequence $\{..., -2, -1, 0, 1, 2, ...\}$ from now on.

Having defined the elements of $\mathbb{Z}$ and the operations of addition and multiplication, Dedekind proceeded to check that these operations inherit the properties stated above for $(\mathbb{N}_0, +, \times)$. Using the simplified notation, we can perform any *addition* (for a, b in $\mathbb{Z}$, $a+b$ is again in $\mathbb{Z}$). The neutral element 0 satisfies $a + 0 = a$ for every a in $\mathbb{Z}$ (adding 0 changes nothing). Every a has an *inverse* for addition: for any $a \in \mathbb{Z}$ there is a *unique* element, which we denote by $-a$, in $\mathbb{Z}$, satisfying the identity $a + (-a) = 0$. This enables us to define *subtraction* in $\mathbb{Z}$ in terms of addition, since $a - b$ is simply shorthand for the addition of a and $-b$, that is: $a - b = a + (-b)$.

In $\mathbb{Z}$ we can *multiply* two numbers and stay within the set. The neutral element is $1 : a \times 1 = a$ for each a in $\mathbb{Z}$.

We extend the *ordering* $a < b$, where the notation means that a is *less than* b if a occurs before b in the ordering. This is again defined to mean that $b = a + k$ for some $k \neq 0$ in $\mathbb{N}$. The 'non-strict' order relation $a \leq b$ again means that either $a < b$ or $a = b$. Both definitions extend those used in $\mathbb{N}$. As usual, we also write $a > b$ if $b < a$ (and $a \geq b$ if $b \leq a$).

As $\mathbb{N}$ is closed under addition, the ordering is *transitive*: if $a < b$ and $b < c$ then $a < c$.

We note (without proof) that we can compare *any* two integers (formally, the ordering is *total*), and that in $\mathbb{Z}$ we have the

(iv) *trichotomy:* Given integers a, b, exactly one of the following three possibilities occurs:

$$a < b, a = b, b < a.$$

We also omit proofs, for $\mathbb{Z}$, of the familiar 'laws of arithmetic', listed as (i)-(iii) for $\mathbb{N}_0$ at the outset, as well as (iv) and the following simple consequences of the above:

The ordering of integers is *compatible* with addition and multiplication. In other words:

(v) if $a < b$ then $a + c < b + c$ for any integer c;

(vi) if $a < b$ then $ac < bc$ if $0 < c$, and $bc < ac$ if $c < 0$

(the final inequality follows as multiplication by -1 reverses the order).

Moreover, the familiar *cancellation laws* hold:

(vii) if $a + b = a + c$ then $b = c$;

(viii) if $a \neq 0$ and $ab = ac$ then $b = c$.

The first cancellation law confirms that *subtraction* (the opposite of addition) gives a unique answer:

if $b + x = a$ and $b + y = a$ then $x = y$.

The use of all these properties will be familar from school mathematics.

Given two pairs from $\mathbb{N}_0$, (m, n) and (p, q), representing *negative* integers, so that $m < n$ and $p < q$, their product $(mp + nq, mq + np)$ will have its first term larger than its second. To see this, we use laws (i)-(iii) and (vi), and compute

$$(mp + nq) - (mq + np) = m(p - q) + n(q - p) = (n - m)(q - p) > 0.$$

In particular, this proves the claim (already used by *Diophantus,* see **Chapter 1**) that $(-1) \times (-1) = 1$.

In the Appendix to **Chapter 8** the above laws will be used to verify our earlier claim that the system of complex numbers $(\mathbb{C}, +, \times)$ *cannot* be given an ordering that is similarly compatible with addition and multiplication.

3.2. Dedekind on rational numbers. Fitting fractions into an extended number system in which the laws of arithmetic remain valid also requires care. However, Dedekind's solution of this issue may feel somewhat more familiar than the steps he needed to accommodate negative numbers. During the Renaissance, as we saw, the arithmetic of fractions became accepted earlier and with less hesitancy than did negative numbers. The practice of expressing a fraction $\frac{m}{n}$ 'in lowest form' is familiar from school. We now express this by saying that we require m and n to be *relatively prime*. The approach taken by Dedekind below simply expresses this practice more formally.

To define the number system $(\mathbb{Q}, +, \times, <)$ formally, Dedekind was able to start with *pairs* (a, b) *of integers,* using the arithmetic and order structure of $(\mathbb{Z}, +, \times, <)$, just as he had used $\mathbb{N}_0$ when defining the integers. To arrive at the integers as a system closed under subtraction, whose elements were defined uniquely, he had needed the equivalence relation discussed above. His task now would be to accommodate another arithmetical operation, division, that was missing from $\mathbb{Z}$, and to define the elements of the new set $\mathbb{Q}$ uniquely. These constraints are again met by defining a suitable choice of equivalence relation, one that is based on what is familiar to us as cross-multiplication.

Dedekind began with the Cartesian product $\mathbb{Z} \times \mathbb{Z}$. He then discarded all pairs whose second coordinate is zero, and imposed an equivalence relation on the remaining pairs:

If integers b and d are not 0, then (a, b) is related to (c, d) whenever $ad = bc$.

This relation is written as $(a, b) \sim (c, d)$.

As we did for addition above, it is easy to check that $\sim$ is reflexive, symmetric and transitive (see Footnote 4). For this we use the fact that multiplication in $\mathbb{Z}$ is commutative and associative, and apply the second cancellation law. Moreover, for any a, b we have $a(-b) = b(-a)$. Thus $(a, b) \sim (-a, -b)$. So in any equivalence class we can find a representative pair whose second element is in $\mathbb{N}$.

In this way, Dedekind defined the set $\mathbb{Q}$ of *rational numbers* as the collection of all *equivalence classes* of pairs of integers (a, b) with $b \neq 0$. As Hamilton had done for the complex numbers, he needed to define the operations of addition and multiplication in $\mathbb{Q}$, ensuring that these reflected the familiar arithmetic of fractions: for given fractions $\frac{m}{n}, \frac{p}{q}$,

$$\frac{m}{n} + \frac{p}{q} = \frac{mq + np}{nq}, \quad \frac{m}{n} \times \frac{p}{q} = \frac{mp}{nq}.$$

Writing $[a, b]$ for the equivalence class of the pair (a, b), Dedekind therefore defined the *sum* of the two classes $[a, b]$ and $[c, d]$ as the equivalence class of the pair $(ad + bc, bd)$:

$$[a, b] + [c, d] = [ad + bc, bd],$$

and their *product* as:

$$[a, b] \times [c, d] = [ac, bd].$$

Recall that the representatives can be chosen with b, d as natural numbers (i.e. positive integers). As we saw for the integers, the classes on the right-hand side *define* the symbols $+, \times$ on the left, as addition and multiplication for rationals.

The resulting rules for the arithmetic of fractions can be derived from these definitions, and we will now revert to denoting the class $[a, b]$ (a rational number) by $\frac{a}{b}$, where $b > 0$ and a and b are relatively prime, to represent this equivalence class of fractions.

The algebraic operations in $\mathbb{Q}$ are compatible with addition and multiplication in $\mathbb{Z}$, which we identify with the subset of $\mathbb{Q}$ where the denominators (second members of the pairs) are 1. Thus $\mathbb{Z}$ may be regarded as a subset of $\mathbb{Q}$, with the same algebraic operations, exactly as $\mathbb{N}$ and $\mathbb{N}_0$ are treated as subsets of $\mathbb{Z}$.

For the ordering of members of $\mathbb{Q}$ simply write: $[a, b] < [c, d]$ if and only if $ad < bc$, where the latter inequality uses the ordering in $\mathbb{Z}$. Here it is important that b and d are taken to be positive. In terms of fractions, this reads: $\frac{m}{n} < \frac{p}{q}$ if and only if $mq < np$, extending the ordering from $\mathbb{Z}$ to $\mathbb{Q}$. The trichotomy and the transitivity of the ordering also extend to $\mathbb{Q}$.

Having defined $(\mathbb{Q}, +, \times, <)$, Dedekind had now arrived at a system of numbers in which addition and multiplication can be performed without restriction or moving outside $\mathbb{Q}$ (in formal terms, the set $\mathbb{Q}$ is *closed* under these operations). The laws of arithmetic remain true in the number system $(\mathbb{Q}, +, \times)$. We can reverse the arithmetical operations: rational numbers can be subtracted at will, and a rational number can be divided by any rational number *other than* 0, with the unique answer remaining an element of $\mathbb{Q}$. Finally, the inherited ordering on $\mathbb{Q}$ is compatible with the arithmetical operations on $\mathbb{Q}$, exactly as described above (in (v) and (vi)) for $\mathbb{Z}$.

However, unlike $\mathbb{N}_0$, the larger sets $\mathbb{Z}$, $\mathbb{Q}$ do not have the Well-Ordering property (although it holds for subsets of $\mathbb{Z}$ that are bounded below). Clearly, neither $\mathbb{Z}$ nor $\mathbb{Q}$ has a smallest element.

Moreover, in $\mathbb{N}$ we identify $n + 1$ as the *successor* of the natural number n. This concept extends to members of $\mathbb{Z}$, but it no longer makes sense in $\mathbb{Q}$: given a rational number r, there is no such thing as the 'next greatest' rational number. If such a number (say s) were to exist, it would obviously satisfy $s > r$, so that $s - r > 0$. But then their average $\frac{r+s}{2}$ would also be rational, and greater than r, but less than s. So the assumption that such an s exists leads to a contradiction: hence there is no such s. Similarly, $\mathbb{Q}^+$, the set of all positive rationals, can have no least member.

In fact, given any two rational numbers $r < s$ one can always find rationals (such as $\frac{r+s}{2}$) that lie strictly between them. Repeating this with r and $\frac{r+s}{2}$ (or, alternatively, $\frac{r+s}{2}$ and s) and continuing in this vein it becomes possible to insert an *infinite* number of *distinct* rationals between r and s, however close together r and s may be. This shows that $\mathbb{Q}$ is very different from the set of natural numbers $\mathbb{N}$—one can 'pack in' rational numbers on the 'number line' as closely as one pleases.

The set $\mathbb{N}$ of natural numbers provides the 'spine' on which the other numbers systems are built.

The well-known *Archimedean property* holds in $\mathbb{Q}$:

Given any p in $\mathbb{Q}$ there is a natural number $n > p$.

Archimedes uses an equivalent statement in his treatise *On Sphere and Cylinder*. However, calling it the 'Archimedean property' is really a misnomer; for line segments, such a statement already appears and plays a key role in the comprehensive study of incommensurable geometrical magnitudes that dominates Book X of Euclid's *Elements*. This book was written before Archimedes was born.

For $\mathbb{Q}$ the proof of the claim is trivial: since $p \in \mathbb{Q}$ has (a representative of) the form $\frac{a}{b}$, where a is an integer and b a natural number, the claim is equivalent to saying that we can find a natural number n such that $nb > a$. This is obvious if $a \leq 0$, while if $a \geq 1$, we can take $n = a + 1$. Then, as $b \geq 1$, we have $nb = (a+1)b = ab + b \geq a + b > a$.

Although all the above properties of $\mathbb{Q}$ had been tacitly assumed to hold for centuries, Dedekind had shown that one can derive this number system by logical reasoning alone, purely on the basis of properties of his starting point, the natural numbers. Just as Hamilton had done for 'imaginary' numbers, he was able to confirm the validity of the inherited arithmetical relationships of his number system, without needing to concern himself with explaining the nature of either negative numbers or fractions in terms of analogies with geometry or anything else. This approach differs sharply from earlier concerns about whether negative or imaginary numbers exist— or, indeed, whether ratios should be regarded as numbers. More important than the nature of the objects were the *relationships* between them that governed their interaction. In time, this *abstract* approach came to govern much thinking about mathematics. In the next section, we will see how it also led Dedekind to a rigorous arithmetical description of the 'number line'.

4. Dedekind cuts

In 1858, when preparing an introductory lecture course on Calculus, Dedekind realised that he could not prove results such as the IVT without an arithmetical definition of the underlying continuum. As he points out in [8], one could not give a rigorous proof of much simpler results, such as $\sqrt{2} \times \sqrt{3} = \sqrt{6}$, unless square roots were described by arithmetical means.[6]

In keeping with his general programme of creating each extended number system from the previous one, Dedekind set out to define a new number

---

[6]In the next chapter we discuss more fully why specific concerns about the geometric description of irrationals were becoming more widespread by the middle of the nineteenth century.

system, starting from $\mathbb{Q}$, by means analogous to the extensions described in the previous section. While those extensions included abstract notions such a equivalence relations, neither had been technically difficult, and in each case the relationship between a number in the extended class and its antecedents in the former one was clear.

However, his current extension, beginning with the rationals, presented a more fundamental obstacle, as he discusses in detail in [8]. He realised in particular that, instead of using pairs of rationals, he would require an *infinite* number of rationals when filling a 'gap' in $\mathbb{Q}$ with an arithmetically defined irrational number, in order that the new number system would have a new property, that he would call *continuity*. His project led him to consider sets of rational numbers more abstractly, and his final definition of elements of the real number system may seem some way removed from the general reader's intuition of what numbers are. The brief summary we give below may therefore be somewhat more challenging conceptually than what has gone before. (More details are given in *MM*.)

Dedekind's definition of real numbers as produced by what he called *cuts* of the set $\mathbb{Q}$ is today one of the standard ways of introducing the real numbers. Although completed in 1858, he did not publish his work for more than a decade. He only did so in 1872 when he became aware that a paper by Georg Cantor, defining real numbers in terms of Cauchy sequences of rationals, was soon to appear. Their papers were by no means universally appreciated at the time. As an example, we have the verdict of the outspoken *Hermann Hankel* (1839-1873):

Every attempt to treat the irrational numbers formally and without the concept of [geometric] *magnitude must lead to the most abstruse and troublesome artificialities, which, even if they can be carried through with complete rigour, as we have every right to doubt, do not have a higher scientific value.*

But in due course the constructions by Dedekind and Cantor were to become staples of undergraduate mathematics in the twentieth century – even if not always welcomed by that audience, either!

Dedekind's elegant paper ([8]) is entitled *Stetigkeit und Irrationale Zahlen* (Continuity and irrational numbers). He lays great emphasis on continuity, which he regards as the key concept through which he links his reasoning explicitly to the properties of the geometric number line.

He compares the order properties of $\mathbb{Q}$ with the positioning of points on a line L:

1. For rational numbers, if $b < a < c$, we say that *a lies between b* and *c*; just as a point p on L lies between points q and r if r is to its right and q to its left.

2. Between any two distinct rational numbers a, b there are infinitely many other rational numbers; similarly, L contains infinitely many points between distinct points p, q.

3. (a) Fix a rational number a. Split $\mathbb{Q}\backslash\{a\}$ into two classes, A_1 and A_2, where A_1 contains all rational numbers $b < a$, while A_2 contains all rational numbers $c > a$. This leaves us with choosing where to place a. Placing a in A_1 would make a the largest number of A_1; placing a in A_2 would make a the smallest number of A_2. Either choice will ensure that each number in A_1 is less than every number in A_2. (Dedekind's choice will become clear below.)

(b) Fix a point p on the line L. Cut L into two pieces and place every point to the left of p into the line segment P_1 and every point to the right of p into the line segment P_2. We can either include p in P_1 as its right-most point, or in P_2 as its left-most point. Either choice ensures that every point of P_1 lies to the left of every point of P_2.

Thus, to ensure that rational numbers correspond uniquely to the 'splits' of $\mathbb{Q}$ described in 3(a) requires a decision whether A_1 should have a largest element or A_2 a smallest. The possibility that A_1 has greatest element b while A_2 also has a least element c can be ruled out: in that case $b < c$ and $\frac{1}{2}(b+c)$ would belong to neither A_1 nor A_2 since it is larger than b and smaller than c.

As Dedekind points out, to each number in $\mathbb{Q}$ there corresponds one and only one point of L. However, there is a crucial difference between $\mathbb{Q}$ and L, since L contains points describing 'incommensurable lengths' such as $\sqrt{2}$:

'Of the greatest importance, however, is the fact that in the straight line L there are infinitely many points which correspond to no rational number...The straight line L is infinitely richer in point-individuals than the domain $\mathbb{Q}$ of rational number-individuals.

If now, as we desire, we try to follow up arithmetically all phenomena in the straight line, the domain of rational numbers is insufficient...and it becomes absolutely necessary that...[$\mathbb{Q}$]...be essentially improved by the creation of new numbers such that the domain of numbers shall gain the same completeness, or as we may say at once, the same continuity, as the straight line.'

His goal in providing a new description of the real numbers is that *'arithmetic shall be developed out of itself'*, mirroring the way that *'negative and fractional numbers are formed by a new creation'*. His method of creating this richer number system rests on understanding what the continuity of the line L means. His answer is that this lies in the *converse* of the 'splitting' of L:

'If the points of the straight line fall into two classes such that every point of the first class lies to the left of every point of the second class, then there exists one

and only one point which produces this division of all points into two classes, this severing of the straight line into two portions.'

He does not claim that he can *prove* this (seemingly obvious) assertion, but instead takes it as an assertion about the nature of the line L that will be taken as an unproven property. The task he now faces is to create a *number system* that 'completes' $\mathbb{Q}$ by satisfying this continuity property.

4.1. Cuts and order properties. The above comparison of $\mathbb{Q}$ and the line L shows that 'splitting' $\mathbb{Q}$ at a rational number (such as 2) will provide a pair of (A_1, A_2) of disjoint sets that together make up all of $\mathbb{Q}$, such as $A_1 = \{r \in \mathbb{Q} : r < 2\}, A_2 = \{r \in \mathbb{Q} : r \geq 2\}$, where the number 2 provides the least element of A_2. But there are also quite different splittings of $\mathbb{Q}$: for example, we might place all negative rationals as well as all other rationals whose square is less than 2 in A_1, while A_2 consists of the rest, i.e. the positive rationals whose square is greater than 2 (there is no rational whose square equals 2). In symbols:

$$A_1 = \{r \in \mathbb{Q} : r < 0\} \cup \{r \in \mathbb{Q} : r \geq 0, r^2 < 2\}, \quad A_2 = \{r \in \mathbb{Q} : r \geq 0, r^2 > 2\}.$$

Here A_1 has no largest element and A_2 no smallest.[7] This differs fundamentally from splitting the line L into two pieces, and occurs precisely because $\mathbb{Q}$ cannot fill the 'gap' at this point. Dedekind's solution is to turn this fact into a definition – for him:

a (Dedekind) *cut* of $\mathbb{Q}$ is a disjoint pair of non-empty sets (A_1, A_2) making up all of $\mathbb{Q}$ (that is, $A_1 \cup A_2 = \mathbb{Q}$ and $A_1 \cap A_2 = \varnothing$), *and* such that A_1 has no largest element.

Following Landau, we will call A_1 the set of *lower numbers*, and A_2 the set of *upper numbers* for this cut.

When cutting $\mathbb{Q}$ at a *rational* number r, Dedekind's choice ensures that the lower numbers have no largest element, while r becomes the smallest upper number for this cut. For any given cut, every lower number is less than every upper number.

To simplify the notation and terminology, we can focus on the set A_1 of lower numbers, since this determines A_2 completely as its *complement* $A_1^c = \mathbb{Q} \setminus A_1$. Writing α instead of (A_1, A_2) to denote the cut means that $\alpha = A_1$ and $\alpha^c = A_2$. We use Greek letters to denote cuts; the *elements* of any cut are rational numbers, denoted by letters such as p, q, r, etc. We re-state the definition in these terms:

Definition:

Call a subset α of $\mathbb{Q}$ a *cut* if

(i) α and $\alpha^c = \mathbb{Q} \setminus \alpha$ are both non-empty sets of rationals,

---

[7]The elementary calculations to prove this are left to the reader – they can be found (e.g.) on p.3 of [40], and in *MM*.

(ii) for rationals p, q, if $p \in \alpha$ and $q < p$ then $q \in \alpha$,

(iii) α has no largest element: given $q \in \alpha$, we can find $p \in \alpha$ with $q < p$.

Thus α consists of the *lower* numbers for the cut, and α^c of its *upper* numbers.[8] Two cuts α, β are said to be *equal* if they contain the same rational numbers. We write $\alpha = \beta$. The set $\mathbb{R}$ of all cuts comprises the *real* numbers; each real number is determined uniquely by a subset of $\mathbb{Q}$, the set of its lower numbers.

The ordering of cuts $\mathbb{R}$ is defined by set inclusion.

Definition:

Given two cuts α, β write $\alpha < \beta$ if α is a proper subset of β i.e. $\alpha \subset \beta$.

Hence β contains a lower number that is an upper number for α.

(i) This definition shows that the ordering is *transitive*: given cuts α, β, γ, if $\alpha < \beta$ and $\beta < \gamma$, then $\alpha < \gamma$.

This follows as β has a lower number r that is an upper number for α, while γ has a lower number s that is an upper number for β. But $r < s$ (for β, any upper number s is greater than any lower number). So $\alpha < \gamma$, since s is a lower number for γ but an upper number for α.

(ii) The *trichotomy* holds for the ordering $<$.

For any cuts α, β, exactly one of $\alpha < \beta$, $\alpha = \beta$, $\beta < \alpha$ holds.

This is immediate from set inclusion, which defines $<$.

We write $\alpha \le \beta$ if either $\alpha = \beta$ or $\alpha < \beta$. Thus $\alpha \le \beta$ means that $\alpha \subseteq \beta$.

As for any (linearly) ordered set, we can define the following useful objects related to the order $\le$ on $\mathbb{R}$:

given a set B of cuts, the cut α is an *upper bound* for B if $\beta \le \alpha$ for all β in B. We say B is *bounded above* by α.

The *least upper bound* (or *supremum*) of B is defined as an upper bound α of B such that $\alpha \le \gamma$ for every upper bound γ of B.

Similarly, a cut δ is a *lower bound* for B if $\delta \le \beta$ for all β in B, and then B is *bounded below* by δ.

The *greatest lower bound* (or *infimum*) of B is defined as a lower bound δ of B such that $\gamma \le \delta$ for every lower bound of B.

The crucial property of *completeness* (or 'lack of gaps') of the set of cuts is then given by the following result.

Theorem:

---

[8]Dedekind's choice, expressed as requirement (iii), imposes an equivalence relation on the set of all pairs (A_1, A_2) of disjoint subsets of $\mathbb{Q}$ that satisfy $A_1 \cup A_2 = \mathbb{Q}$. A cut is an equivalence class of such pairs, represented by its set of lower numbers.

Let B be a non-empty subset of $\mathbb{R}$ for which there exists an upper bound $\theta \in \mathbb{R}$ under the ordering $<$. Then B has a least upper bound in $\mathbb{R}$.

Proof: Each $\alpha \in B$ is a non-empty subset of $\mathbb{Q}$. Let $L(B)$ be the union of all the sets of *lower numbers* for cuts in B. Thus $r \in \mathbb{Q}$ belongs to $L(B)$ if and only if there is at least one cut $\alpha \in B$ such that r is a lower number for α. A rational number r may be a lower number for many cuts in B, but we list it only once in forming $L(B)$. It is a straightforward exercise to show $L(B)$ satisfies the definition of a cut (see *MM* for details).

Next, $L(B)$ is an upper bound for B since, by definition of $L(B)$, the set of lower numbers for any α in B is a subset of $L(B)$. So $\alpha \leq L(B)$ holds for every $\alpha \in B$ by definition of the ordering (since $\leq$ means $\subseteq$).

Finally, $L(B)$ is the *least* upper bound for B: if a cut η is an upper bound for B we must show that $L(B) \leq \eta$. So we need to show that every lower number for $L(B)$ is also a lower number for η. But the lower numbers for every α in B are lower numbers for η (as η is an upper bound for B). It follows that $L(B)$ is a subset of the set of lower numbers for η, i.e. $L(B) \leq \eta$. Hence $L(B)$ is the least upper bound for B. This completes the proof.

Rational cuts are easily identified:

Theorem

Given a rational p, define $\alpha_p = \{r \in Q : r < p\}$. Then α_p is a cut, and its smallest upper number is p.

Proof: Clearly α_p satisfies (i), (ii) of the definition of a cut. To see that α_p has no greatest lower number, let $q \in \alpha_p$ be arbitrary, so that $q < p$. But then $q < \frac{1}{2}(q+p) < p$ and (by definition of α_p) we have $\frac{1}{2}(q+p) \in \alpha_p$, so (iii) is satisfied and α_p is a cut.

Equally, we cannot have $p < p$, so p is not in α_p, hence it is an upper number for α_p. But $q < p$ implies $q \in \alpha_p$, so any upper number q must satisfy $q \geq p$, hence p is the smallest upper number for α_p.

We call α_p the *rational cut* associated with p. We have a one-one correspondence $p \leftrightarrow \alpha_p = \{r \in \mathbb{Q} : r < p\}$ between elements of $\mathbb{Q}$ and elements of $\mathbb{R}$. The correspondence preserves the ordering of $\mathbb{Q}$: $p < q$ if and only if

$$\{r \in \mathbb{Q} : r < p\} \subset \{r \in \mathbb{Q} : r < q\}.$$

This indicates how the set $\mathbb{Q}$ can be embedded in the set $\mathbb{R}$ in a way that preserves the ordering. Moreover, it is not hard to show that between any two given cuts there is a rational cut.

Defining the sum and product of two cuts and checking that the arithmetical properties correspond takes somewhat more work, but clearly the rational cut $\{r \in \mathbb{Q} : r < 0\}$ is the neutral element for addition. This enables Dedekind to define positive and negative cuts, and check that the laws of arithmetic extend from $\mathbb{Q}$ to $\mathbb{R}$. Proofs of these claims can be found in *MM* and in (e.g.) [8], [10], [40]. (Also see [6]).

The steps outlined here enabled Dedekind to characterise real numbers directly in terms of rationals, showing that (as suspected) the real number α can be regarded as the set of all rationals less than α. Although he avoided the logical trap that Cauchy fell into when defining irrationals as limits of rational sequences, Dedekind himself was somewhat hesitant about explicitly calling cuts numbers: for him, the set of lower numbers of a cut 'produces' a real number. Today, mathematicians generally do not share these qualms. A football analogy may be helpful here: when Manchester United play against Liverpool, the pitch will usually feature eleven players in the set comprising the Liverpool team, yet when the result of the match is recorded in the League Table, the team is regarded as a single unit (Liverpool). The *set* of players is treated as a single *element* of another set, namely the set of Premier League teams. Similarly, a cut is a set of rationals, but this set is equally regarded as an element of the set of real numbers.

The real number system that Dedekind defined fills in all the gaps of $\mathbb{Q}$. The question now arose whether *repeating* the process would again generate new elements if one took cuts of the reals, as was done for the rationals. Dedekind's key result shows that this would not happen—the analogy with the line had been 'completed' by defining the cuts.

Dedekind's Theorem:

Given non-empty subsets A, B of $\mathbb{R}$ such that $A \cup B = \mathbb{R}$, $A \cap B = \phi$ and $a \in A$, $b \in B$ implies $a < b$, there is a unique $x \in \mathbb{R}$ such that for all $a \in A, b \in B$ we have $a \leq x \leq b$.

Consequently, either A has a largest element or B a smallest.

The proof can be found in *MM*.

5. Cantor's construction of the reals

Georg Cantor expressed his admiration for the elegance and clear logic of Dedekind's construction, but he was adamant that *'numbers in analysis never present themselves as "cuts," and therefore have first of all to be brought into this form by elaborate artifices'.* His own construction, which he felt was the *'most natural of all'*, presenting the approach best suited to the analysis of functions, started from the *Cauchy criterion* (see **Chapter 6**) for convergence of sequences. In 1872 Cantor outlined his ideas in a paper dealing with trigonometric series, which he had sent to Dedekind prior to publication, thus prompting Dedekind to publish his long-held views. His was not the first publication using Cauchy sequences to define the irrationals—in 1869, the French mathematician *Charles Meray* (1835-1911) had published an account in different terminology harking back to Cauchy. *Eduard Heine* (1821-1881), Cantor's colleague at the University of Halle, who was motivated more directly by the need for clear foundations for Real Analysis, also published a version of Cantor's arguments in a paper in 1872.

Cantor's initial motivation differed substantially from that of other writers on the subject. His main interest, first encouraged by Heine, was to explore the way *Fourier series* were able to represent a wide class of functions.[9] Cantor addressed the problem of deciding for which functions the representing series is uniquely defined. He showed initially that this will be true if the series converges to f at all points. By 1872, he had succeeded in proving uniqueness even if the convergence failed at *infinitely many* points, provided that these points are distributed on the line in a specific fashion. It was this result that led him to consider how rational and irrational 'points on the line' are distributed, and thus drew his attention to the need for irrationals to be defined unambiguously as numbers.

Cantor was well aware that he could not define a real number as the limit of a sequence of rationals, since the definition of limit involves identifying the limit as a number. *Cauchy sequences* of rationals, however, only require us to know what rationals are and how to do arithmetic with them. Moreover, any rational number r has an obvious associated Cauchy sequence, namely the infinite constant sequence $\{r, r, r, r, ..., r...\}$. On the other hand, Cauchy sequences such as our old friend, the successive decimal approximations $\{1, 1.4, 1.41, 1.414, ...\}$ to $\sqrt{2}$, clearly lead to irrationals.

Cantor's definition of irrationals was discursive, leaving many details to the reader. He considered Cauchy sequences of rationals, or, as he put it, 'fundamental sequences of the first order'. To each such sequence, consisting of rationals (a_ν) such that *'after the choice of an arbitrarily small rational number ε a finite number of members can be separated off, so that those remaining have pairwise a difference which in absolute terms is smaller than ε'*, he attached a symbol b. The *real numbers* would constitute the collection of all such first-order sequences. Cantor was aware that different Cauchy sequences can have the same limit, but it was his colleague Heine, who, in a paper also

---

[9]This had been a hot topic in analysis for some decades, ever since *Joseph Fourier's* famous 1807 investigation on representing a general function f on the interval $[-\pi, \pi]$ by a trigonometric series of the form $\frac{1}{2}a_0 + \sum_{n=1}^{\infty}[a_n \cos(nx) + b_n \sin(nx)]$. Fourier had calculated the coefficients in the series expansion as $a_n = \frac{1}{\pi}\int_{-\pi}^{\pi} f(t)\cos(nt)dt$, $b_n = \frac{1}{\pi}\int_{-\pi}^{\pi} f(t)\sin(nt)dt$, and claimed that, with these coefficients, the series would converge at each point x to the value $f(x)$. The counter-example given by Niels Abel to Cauchy's assertion that an infinite sum of continuous functions is always continuous used such a sequence. Abel also realised that his result illustrated deficiencies in various proofs given by Fourier. Over the next several decades a number of prominent mathematicians, including Dirichlet and Heine, had gradually succeeded in widening the range of situations for which uniqueness of the representation of f could be proved.

Figure 36. Georg Cantor, by an unknown photographer, ca. 1900[10]

published around the same time, clarified explicitly that two such sequences should be considered equal if the sequence of their differences converges to 0. This requirement again introduces an *equivalence relation* on the set of rational Cauchy sequences. The fundamental sequences $(a_n), (b_n)$ belong to the same equivalence class if $|a_n - b_n| \to 0$ when $n \to \infty$.

Cantor was at pains to stress that b was in no sense assumed to be a 'limit', but served simply as a symbol representing the sequence. It was only after carefully defining the algebraic and order relations for fundamental sequences that he could make sense of the statement $b = \lim_{\nu \to \infty} a_\nu$. He argued that irrationals, as a result of their definition, should have *'as definite a reality in our mind'* as do rationals, and that this was what should convince one of the *'evident admissibility of the limiting processes'*. He then used the same construction, taking fundamental sequences of the second order, where each element was a fundamental sequence of the first order, to create the next level of abstraction. He continued this process indefinitely, but showed that *nothing new* would be created in any such repetition: they all *'accomplish exactly the same thing for the determination of a real number b as the fundamental sequences of the first order'*. This became his version of Dedekind's 'continuity' property.

We summarise Cantor's approach in current terminology:

Definition

---

[10]https://commons.wikimedia.org/wiki/File:Georg_Cantor_(Porträt).jpg

(i) A *Cauchy* (or *fundamental*) sequence of rational numbers $(a_n)_{n \geq 1}$ is such that, for any given rational $r > 0$, there is a natural number N, depending on r, such that $|a_m - a_n| < r$ whenever $m, n \geq N$. Denote the set of all rational Cauchy sequences by $\mathcal{C}$.

(ii) A *null* sequence $(a_n)_n$ of rational numbers is such that, for any rational $r > 0$, there is a natural number N, depending on r, such that $|a_n| < r$ for all $n \geq N$. Denote the set of all rational null sequences by $\mathcal{N}$.

(iii) Two Cauchy sequences $(a_n), (b_n)$ of rational numbers are *equivalent* if the sequence $(a_n - b_n)_n$ of their differences is a null sequence. Write this as $(a_n) \sim (b_n)$, or $(b_n) \in (a_n) + \mathcal{N}$. This defines an equivalence relation on $\mathcal{C}$.

Definition: The set $\mathbb{R}$ of real numbers is the set of all *equivalence classes* of Cauchy sequences of rationals under the relation $\sim$. In other words, for any sequence $\mathbf{s} = (s_n)$ in $\mathcal{C}$, the equivalence class of $\mathbf{s}$ is the subset of $\mathcal{C}$ given by

$$[\mathbf{s}] = \{\mathbf{s}' \in \mathcal{C} : \mathbf{s}' \sim \mathbf{s}\}.$$

Any constant sequence $\mathbf{r}$ with $r_n = r$ for all n is in $\mathcal{C}$. For two constant sequences the equivalence $\mathbf{r} \sim \mathbf{s}$ means that $r = s$. Thus $[\mathbf{r}]$ is the only constant sequence in its class. We identify the rationals with the equivalence classes of constant sequences, i.e. *embed* $\mathbb{Q}$ in $\mathbb{R}$ by the correspondence $r \leftrightarrow [\mathbf{r}]$. So the rational number r is represented in $\mathbb{R}$ by the class of the constant sequence $\mathbf{r} = \{r, r, r, r, ..., r, ...\}$.

To define the algebraic operations for the set of all equivalence classes, addition and multiplication of the representing sequences is done coordinate-wise – that is, individually for corresponding terms in the sequence. For given rational Cauchy sequences $\mathbf{x} = (x_n)_n$ and $\mathbf{y} = (y_n)_n$, the *sum* $[\mathbf{x}] + [\mathbf{y}]$ becomes the class of the sequence $\mathbf{x} + \mathbf{y} = (x_n + y_n)_n$ and similarly for their product. The class $[\mathbf{0}]$ of the constant sequence with $r = 0$ is the neutral element for addition: it consists of all null sequences $(c_n)_n$ in $\mathcal{C}$. Hence $[\mathbf{a}] + [\mathbf{0}] = [\mathbf{a}]$ for all classes $\mathbf{a}$. Similarly, the class $[\mathbf{1}]$ is the neutral element for multiplication. Inverses are also simple to determine – note that to have a multiplicative inverse, the class $[\mathbf{a}]$ must not contain null sequences.

This defines the algebraic system $(\mathbb{R}, +, \times)$, in which the rationals $\mathbb{Q}$ are embedded by the correspondence $r \leftrightarrow [\mathbf{r}]$. We may now treat $\mathbb{Q}$ as a subset of $\mathbb{R}$, with $+$ and $\times$ extended from $\mathbb{Q}$ to $\mathbb{R}$.

This embedding allows Cantor to extend the *modulus* $|\cdot|$ to $\mathbb{R}$ by taking the modulus $||\mathbf{x}||$ of $\mathbf{x} = (x_n)$ as the class of $(|x_n|)_n$, and to *define* the set of (strictly) *positive* real numbers as

$$\mathbb{R}_+ = \{[\mathbf{x}] \in \mathbb{R} : [\mathbf{x}] \neq [\mathbf{0}], ||\mathbf{x}|| = [\mathbf{x}]\}.$$

A real number $[\mathbf{y}]$ is *negative* if its additive inverse $-[\mathbf{y}]$ is positive. Setting $[\mathbf{0}] < [\mathbf{x}]$ if $[\mathbf{x}]$ belongs to $\mathbb{R}_+$ defines the ordering $<$ for $\mathbb{R}$.

The proof of completeness of $\mathbb{R}$ uses the Cauchy Criterion for convergence – in Cantor's approach this is the more technically challenging argument and will be omitted here. (Proofs can be found in MM and [10].)

Unlike Dedekind, Cantor emphasises the algebraic operations rather than the ordering. This makes his approach suitable for extensions of the completeness property to domains which have no linear ordering, such as the Cartesian plane $\mathbb{R}^2$ and its extensions to higher dimensions.

6. Decimal expansions

Having defined real numbers arithmetically, we can use the familiar concept of *decimal expansions* to highlight the difference between irrationals and rationals in another way that may feel more 'concrete' than Dedekind cuts or classes of Cauchy sequences of rationals. We will concentrate on positive numbers here. In our base-10 positional system of writing numbers, the *decimal expansion* of a real number is obtained by making successive choices from the set of single-digit (base 10) numbers $\{0, 1, 2, 3, 4, 5, 6, 7, 8, 9\}$, and combining them with the decimal point in order to separate the 'integral' and 'fractional' parts. The decimal point indicates where we leave the realm of natural numbers, so that what comes after the decimal point is intended to represent a number 'lying between' 0 and 1.

6.1. Expanding rationals. Write an *infinite decimal expansion* of a positive real number as

$$a_0.a_1a_2...a_m...$$

where a_0 is an element of $\mathbb{N}_0$, while, for $j \geq 1$, each a_j is one of the digits $0, 1, 2, 3, 4, 5, 6, 7, 8$ or 9. Here a_0 represents a finite sum of multiples (using these digits) of positive powers of 10. After the decimal point we add the *infinite* sum

$$\sum_{n=1}^{\infty} \frac{a_n}{10^n} = \frac{a_1}{10} + \frac{a_2}{10^2} + ... + \frac{a_k}{10^k} + ...$$

Finite, or 'terminating', decimal expansions are familiar, of course: we write $\frac{1}{8}$ as 0.125, for example. Nothing really changes if we write this as the *infinite* decimal expansion $0.12500000...$ instead, adding zeroes in all the decimal places after the third. Doing so brings the terminating expansion into line with 'recurring' expansions like $\frac{1}{3} = 0.3333.... = 0.\overline{3}$ or $\frac{1}{7} = 0.\overline{142857}$, where in both cases the upper bar denotes the indefinite repetition of the number or group of numbers concerned. (The recurrence need not start immediately after the decimal point: consider $\frac{8}{15} = 0.53333...$, for example.)

Any positive rational number, represented by $r = \frac{m}{n}$ (where the natural numbers m, n have no common factors) will have decimal expansion of one of these two forms: if n has no prime factors other than 2 and 5, its decimal expansion terminates, because the fraction $\frac{m}{n}$ is equivalent to one of the form $\frac{q}{10^k}$, for some natural number q and $k \geq 1$.

Any other rational will have a non-terminating expansion. If, for example, $\frac{1}{3}$ had a terminating decimal expansion, we would have $\frac{1}{3} = \frac{q}{10^k}$ for some natural numbers q and $k > 1$. Hence $10^k = 3q$. But the uniqueness of prime factorisation shows that these numbers cannot be equal: on the left the prime factors are 2 and 5, while on the right 3 is a prime factor. A similar argument clearly applies to any rational $\frac{m}{n}$ where n contains any prime factor other than 2 or 5.

We characterise recurring expansions of fractions $\frac{m}{n}$ by using the *pigeonhole principle*[a]:

Suppose there are k pigeonholes and $n > k$ pigeons. Then at least one pigeonhole contains more than one pigeon.

For, if *no* pigeonhole contains more than one pigeon, and $l \geq 0$ pigeonholes are unoccupied, then $k - l$ pigeonholes contain exactly one pigeon. The two *finite* sets, respectively of occupied pigeonholes and of pigeons, have the same number of elements, i.e. $k - l = n$. This contradicts the fact that $n > k$. So our assumption is false, and hence at least one pigeonhole contains more than one pigeon.

Now apply this to long division. If the decimal expansion of the rational number $\frac{m}{n}$ is infinite, then dividing m by n must leave a non-zero remainder at infinitely many decimal places, as the expansion does not terminate. These remainders must be natural numbers less than n, which means that there are at most $(n-1)$ different choices. Hence within n successive decimal places some remainder r will occur a second time. As we are dividing $m.000....$ by n, the immediate successor of r will repeat exactly as before. In other words, the decimal expansion of any rational number either terminates or recurs with a finite period.

[a]Also known as *Dirichlet's* 'drawer principle' [Schubfachprinzip]. He used it in 1834, but its first known appearance is in a 1624 text by the French Jesuit priest *Jean Leurechon*.

We combine terminating and recurring expansions in the single phrase *eventually periodic* (with periodicity starting after a finite number $k \geq 0$ of places after the decimal point). We have shown that every rational number has an eventually periodic decimal expansion.

The converse is also true: if the decimal expansion of a real number x is eventually periodic, then x is rational.

To see this, we may assume that the periodicity begins at the first decimal place (if the expansion begins with m zeroes after the decimal point, we consider $10^m x$ instead). Thus $x = a_0.\overline{a_1 a_2 ... a_k}$. This is the sum S of an infinite series with first term a_0. Let $\sum_{n=0}^{\infty}$ denote summation of all its terms, starting at $n = 0$, and set $q =$

$\frac{a_1}{10} + \frac{a_2}{10^2} + ... + \frac{a_k}{10^k}$, then S is the rational number

$$S = a_0 + q\sum_{n=1}^{\infty}\frac{1}{(10^k)^n} = a_0 + q\left[\frac{\frac{1}{10^k}}{1 - \left(\frac{1}{10^k}\right)}\right].$$

Therefore the rational numbers are *characterised* as real numbers whose decimal expansions are eventually periodic. To know the infinite decimal expansion of a rational number r *exactly*, it suffices to know *finitely many* of its initial entries after the decimal point.

As we will see shortly, it is often helpful to express any terminating decimal expansion as an infinite expansion, in order to obtain a consistent set of entities. However, when doing this, the need to describe such a rational number *uniquely* as an infinite decimal expansion will require us to make a choice between using recurring nines or recurring zeroes. Like Dedekind did for his cuts, we will opt for the former, as will be shown shortly.

6.2. Expanding irrationals. Call an infinite decimal expansion *aperiodic* if it is not eventually periodic. Thus the *irrational* numbers can be characterised as real numbers whose decimal expansions are aperiodic. For these expansions we *cannot* determine the full expansion after only seeing finitely many decimal entries, since at each stage we have no *a priori* means of determining the next digit with certainty. An infinite sequence of digits chosen at random is highly likely to be aperiodic.

Imagine dipping a fishing net into a pond containing all infinite decimal expansions, and pulling out one of them, x, at random. For x to be eventually periodic, it must have the property that there exists some finite M such that, having read the first M digits of $x = a_0.a_1a_2a_3...$, we will, from that point onward, know for certain what *all* its remaining digits are. A randomly selected infinite sequence of decimal digits is highly unlikely to have this property. At each point there are 10 choices for the next digit, which – in the absence of additional information – might reasonably be considered to be equally likely. They are chosen independently of each other. Under these assumptions the occurrence of any particular sequence of k digits in the infinite decimal expansion would have probability $\frac{1}{10^k}$, unless we have extraneous information. This makes it highly implausible, if k is large, that our 'randomly chosen' expansion could be eventually periodic.

This suggests that the infinite decimal expansion of a real number is a more subtle concept than may at first appear to be the case. We will consider this (possibly disturbing) issue further in **Chapter 10**.

6.3. The Weierstrass-Stolz model. Representing a positive real number x by an infinite decimal expansion $a_0.a_1a_2a_3...$ makes perfect sense once the set $\mathbb{R}$ has been defined: for any $m \geq 1$, we call the *finite* decimal expansion of order m, $x(m) = a_0.a_1a_2a_3...a_m$ the mth *truncation* of this expansion (with the proviso, as indicated below, that a terminating expansion is shown in its 'recurring nines' form).

Then $x(m)$ represents the *rational* number

$$x(m) = \frac{10^m a_0 + 10^{m-1} a_1 + ... + a_m}{10^m}.$$

The sequence of truncations $(x(m))_{m \geq 1}$ is non-decreasing (at each decimal place we add a non-negative term) and it is certainly bounded above by $a_0 + 1$. By the completeness of $\mathbb{R}$ this means that the sequence $(x(m))_m$ converges to a real number, namely the supremum of the sequence. Each $x(m)$ is a partial sum of the infinite series defining this supremum and writing $x = a_0.a_1a_2...a_n...$ is a short-hand notation for the sum of the series. In this sense we can now exhibit any irrational as the *limit* of a sequence of rationals, as Cauchy wished to do.

This argument can be turned on its head, providing a more 'concrete' model of real numbers—although the above characterisation of irrationals as aperiodic expansions suggests that infinite decimal expansions are rather less familiar than they seem! The truncations of an arbitrary infinite decimal expansion define *lower numbers* for a Dedekind cut that yields a real number x. Moreover, they will approach x arbitrarily closely, as the series they define converges to it.

As we have observed, for certain *rational* cuts there is ambiguity: for $r = \frac{q}{10^m}$ we can choose the infinite decimal expansion either in the form

$$a_0.a_1a_2...a_m 0000$$

or, alternatively, as

$$a_0.a_1a_2...(a_m - 1)999...$$

since these two series have the same (rational) sum. For example, $\frac{1}{2} = 0.50000...$ or $0.49999....$ Dedekind's choice that the lower numbers of a cut should *not* have a greatest member is reflected by taking the 'recurring nines' expansion whenever this occurs—this choice constitutes an equivalence relation on the set of decimal expansions, just as Dedekind's choice led to equivalence classes of cuts.

Thus, instead of using Dedekind cuts, we could begin with the set of positive infinite decimal expansions as defined above, impose the equivalence relation just described, and mimic the logic of Dedekind's arguments.

First, define the order relation $<$ for infinite decimal expansions as follows: if $a = a_0.a_1a_2...a_n...$ and $b = b_0.b_1b_2...b_n...$, *define* $a < b$ if there is a $k \geq 0$ such that $a_i = b_i$ for all $i < k$ and $a_k < b_k$.

It is easy to see that the trichotomy holds for $<$, and we can define upper bounds as before. The key result is (again) that any set of infinite decimal expansions that is bounded above will have a least upper bound.

In other words, the set of positive infinite decimal expansions has the completeness property. Extending this to all expansions and, in particular, defining the arithmetical operations $(+, \times)$ for negative numbers, is somewhat more intricate.[11] Checking the arithmetical axioms can become tedious, so we will not pursue the matter here—an outline, including an elementary proof of completeness and the definition of arithmetical operations, can be found in *MM*. This approach is sometimes referred to as the *Weierstrass-Stolz* model of the reals—after *Vorlesungen über Allgemeine Arithmetik*, published in 1885 by Weierstrass' former student *Otto Stolz* (1842-1905).

7. Algebraic and constructible numbers

Motivated by the solution of the 'three famous problems' of antiquity (see **Chapter 3**), we can identify various classes of real numbers in a different way. Recall that Plato reports the discovery by *Thaeatetus* that (in our terminology) positive square or cube roots of natural numbers are either natural or irrational numbers.

7.1. Roots. We place this in a more general setting: given any m in $\mathbb{N}$, the positive m^{th} root of a natural number N is the unique positive solution of the equation $x^m = N$, and is written as $x = \sqrt[m]{N}$.

If $x = \sqrt[m]{N}$ is rational, then we can find relatively prime natural numbers a, b such that $x = \frac{a}{b}$. Now solve the equation $(\frac{a}{b})^m = N$, so that $a^m = Nb^m$. Write the unique prime factorisation of b as $b = p_1^{a_1}...p_k^{a_k}$. Then every prime p_i on the right divides b, so it must also divide a^m. Since p_i is prime, it now divides a, using the extension of Euclid's lemma we proved in Section 2. So a and b are relatively prime, yet have every p_i as a common factor, which is impossible unless $b = 1$, so that $N = a^m$. Hence the Fundamental Theorem of Arithmetic immediately generalises Theaetetus' result, proving that:

$\sqrt[m]{N}$ *is irrational unless* $N = a^m$ *for some natural number* a.

This result provides us with an unlimited collection of different irrational numbers. In particular, this collection includes $\sqrt[3]{2}$, which, if it could

---

[11]For consistency this requires us to write the expansion of a *negative* number in an unusual form: what is normally written as -1.75 is now written as -2.25, which will mean $-2 + 0.25$, so that we can read the part after the decimal point as positive. So, for example, -1.76 becomes -2.24, which is less than -2.25 in the ordering as defined here: the initial digits and the first decimal digits are equal, and in the second decimal place 4 is less than 5. (We would also replace these terminating expansions by their 'recurring nines' version.)

be constructed by straightedge and compass, would solve the ancient problem of the duplication of the cube, considered in **Chapter 3.**

7.2. Algebraic numbers. The positive m^{th} root of N arises as the root of the polynomial $x^m - N = 0$. In **Chapter 4,** the Fundamental Theorem of Algebra showed that any polynomial of degree m will have exactly m (not necessarily distinct) roots belonging to the *complex* number system $\mathbb{C}$. This leads to a definition:

Definition:

An *algebraic number* is a solution of a polynomial equation of the form

$$c_m x^m + c_{m-1} x^{m-1} + \dots + c_1 x + c_0 = 0,$$

where the c_i are integers (or, equivalently, are rational numbers).

By the Fundamental Theorem of Algebra, this equation has m solutions in $\mathbb{C}$.

In particular, the rational $r = \frac{a}{b}$ is an algebraic number, since it is the root of the equation $bx + (-a) = 0$.

We denote the set of all *real* algebraic numbers by $\mathbb{A}$. This includes $\mathbb{Q}$, but is considerably larger, as it includes all positive m^{th} roots of natural numbers, for example. Sums and differences of square roots are similarly included. For example, if a, b are natural numbers, then $x = \sqrt{a} + \sqrt{b}$ satisfies $x^2 = (\sqrt{a} + \sqrt{b})^2 = a + b + 2\sqrt{ab}$ and, collecting terms and squaring again, we obtain $(x^2 - (a + b))^2 = 4ab$ which becomes the polynomial

$$x^4 - 2(a + b)x^2 + (a - b)^2 = 0.$$

In fact, the set of real algebraic numbers includes all 23 classes of incommensurables identified (with considerably greater effort) by Euclid in Book X of the *Elements*: the surds he considered all lead to polynomial equations whose degree is a power of 2. So all Euclid's classes of incommensurables belong to the set of real algebraic numbers.

7.3. Constructible numbers. The set of real algebraic numbers includes the solutions of the first two 'famous problems' of antiquity. First, the duplication of the cube involves constructing a length of $\sqrt[3]{2}$, which is a real algebraic number. Second, we saw in **Chapter 3** that the trisection of the general angle, leads to the cubic equation $4z^3 - 3z - c = 0$, where $c = \cos \phi$ for the given angle ϕ. If c is rational, the three complex roots of the cubic are algebraic numbers. But the Intermediate Value Theorem ensures that any cubic has at least one real root, since the polynomial is negative for large negative z and positive for large positive z. In our example, when $\phi = 60°$ we know that $c = \frac{1}{2}$, so the cubic equation becomes $8z^3 - 6z - 1 = 0$, whose roots include the real algebraic number $z = \cos 20°$.

But, although they are algebraic numbers, neither $\cos 20°$ nor $\sqrt[3]{2}$ represent lengths that can be constructed by straightedge and compass alone. In 1837 the French mathematician *Pierre Wantzel* (1818-1848) defined the class of *constructible* numbers as numbers that correspond to lengths of such line segments. Using a method now known as field extensions, Wantzel proved that these are precisely the numbers that can be obtained by repeated use of the four arithmetical operations $(+, -, \times, \div)$ and *square* root extraction.[12] Obtaining a constructible length from a rational length by straightedge and compass therefore implies that this length represents the root of a polynomial equation with rational coefficients *whose degree is a power of* 2.

We know that $\sqrt[3]{2}$ and $\cos 20°$ are roots of cubic equations, and one can prove that they cannot satisfy an equation whose degree is a power of 2. Thus neither is a constructible number.

Therefore, the set $\mathbb{A}$ of real algebraic numbers includes all constructible numbers as a *proper* subset, and the set of constructible numbers includes the rationals as a *proper* subset. All three are of course infinite subsets of the real number system $\mathbb{R}$. We might reasonably ask if there are any irrationals, representable by aperiodic decimal expansions, that are *not* algebraic numbers?

8. Transcendental numbers

The third 'famous problem' of antiquity asked for the side of the square with area equal to that of a given circle—this is often called the *quadrature* of the circle. In the simplest case the circle has unit radius, so that its area is π, and the side of the square we seek is $\sqrt{\pi}$. It turns out that this number, and π itself, are *not* algebraic numbers—and neither is the irrational number Euler had named e, the base of natural logarithms (see **Chapter 5**).

Real numbers that are not algebraic are called *transcendental* – the term was first used in this context by Euler, following Leibniz' choice of this name for curves that Descartes had called 'mechanical'. But at that stage no one had a proof that such numbers exist.

The earliest candidate to be examined closely was the familiar number π. In 1768 the Swiss polymath *Johann Heinrich Lambert* (1728-1777) proved that π is irrational and conjectured (with a sketch plan for a possible proof) that it must be transcendental. Similarly, in 1806, Legendre, who had shown that π^2 is irrational, concluded about π:

---

[12]In the classic text [7] Courant and Robbins provide a fairly detailed description of Wantzel's methods. Wantzel's paper aroused little interest and was largely forgotten for nearly a century (see a recent account by Jesper Lützen in *Historia Mathematica* 36 (2009) pp. 374-394). Also see *MM* for a remarkable result concerning the construction of regular polygons inscribed in a circle, largely proved by Gauss and later completed by Wantzel.

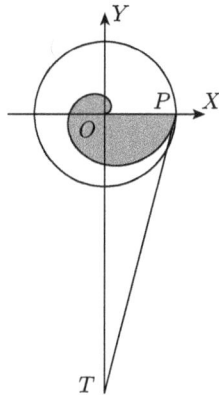

Figure 37. Circle-squaring with the spiral

'It is probable that it is not even contained among the algebraic irrationals; in other words it cannot be the root of an algebraic equation with a finite number of terms, and rational coefficients. However, it seems difficult to prove this theorem rigorously.'

In antiquity, various attempts, some of them ingenious, were made to square the circle.

The *quadratrix* of Hippias, described in **Chapter 3** as an angle trisector, provided an early 'solution' (based on Figure 16(b)) but this curve cannot be constructed by ruler and compass, as Euclid demands. Although no direct written account confirms it, most historians accept the report by the fourth century commentator Pappus that it was Menaechmus' brother *Dinostratus* (about whom not much else is known) who made this discovery around 350 BCE. Pappus was well aware that this does not provide a straightedge-and-compass construction, as the quadratrix cannot be constructed by these means. But his report reflects the name that has survived for this curve, relating it to squaring the circle rather than trisecting the angle.

A different approach, unsuccessful but illustrating the level of sophistication achieved at that time, is preserved in an even earlier fragment by *Hippocrates of Chios*. He was able to square various types of *lunes* (crescent-shaped figures bounded by two circular arcs of unequal radii). He then noted that he would be able to square the circle if he could square the lunes created by a semicircle with half a regular inscribed hexagon and three semi-circles with diameters equal to the sides of the hexagon. However, squaring these particular lunes proved elusive. (See *MM* for details of the above constructions.)

The spiral of Archimedes (see **Chapter 5**) provides a further quadrature of the circle. In Figure 37 the spiral starts at the point *O*, while *P* is the point reached at the first full turn of the circle. The tangent to P meets the line perpendicular to OP at T. In Proposition 19 of his treatise *On Spirals*, Archimedes shows that the line segment OT equals the circumference of the circle OP. If $OP = r$ this means that $OT = 2\pi r$, so that the area of the right-angled triangle TOP equals that of the circle OP. Constructing a square with this area is then a simple matter.

These geometric constructions bring us no closer to a criterion for deciding when a given number must be transcendental, or to define an example of such a number arithmetically. This required a radically different approach, developed in the mid-nineteenth century.

8.1. Rapid approximation by rationals. The breakthrough, showing that transcendental numbers exist, was made in 1844 by the French mathematician *Joseph Liouville* (1809-1882), who was the first to *identify* a specific number that he could prove to be transcendental. His approach was to consider ways in which one might distinguish between the two classes of algebraic and transcendental numbers by considering the *speed* with which such numbers can be approximated by rationals.

It may seem counter-intuitive at first, but what separates the two classes is that transcendental numbers allow rational approximations of *greater rapidity* than algebraic numbers do! To make this statement more precise, consider the approximation of an arbitrary real number by a rational number.[13]

We can find infinitely many rationals between any two distinct real numbers. On the other hand, consider the *distance* between two distinct rational numbers $r = \frac{a}{b}$ and $s = \frac{c}{d}$ (taking b and d as positive). We obtain

$$|r - s| = |\frac{a}{b} - \frac{c}{d}| = \frac{|ad - bc|}{bd} \geq \frac{1}{bd}.$$

The final inequality follows from Dedekind's definition of rational numbers: if we have $ad = bc$, then $r = s$. Thus, for distinct r, s, the distance $|ad - bc|$ is at least 1, as a, b, c, d are integers. In other words, if we try to approximate $s = \frac{c}{d}$ by the rational r, we cannot get closer than the *reciprocal of the product of the denominators* of the two fractions (in lowest form) representing these rationals. We can approximate a fixed rational $s = \frac{c}{d}$ as closely as we please by another rational, but to get very close to s with $r = \frac{a}{b}$, we must expect the denominator b of r to become very large.

Next, suppose that we wish to approximate a given *real* number x by a rational number $r = \frac{a}{b}$ in such a way that the distance between them is less than the square of the denominator of r, that is

$$|x - r| < \frac{1}{b^2}.$$

---

[13]We follow **[18]** pp.159-163, in the arguments given here. The approximation of real numbers by rationals is a much wider topic, discussed in more detail there.

If $x = s = \frac{c}{d}$ is also rational, we have just seen that the distance is at least $\frac{1}{bd}$, which is less than $\frac{1}{b^2}$ only if $b < d$. Hence, if $x = s$ is rational, there can only be $(d-1)$ choices for the denominator b of the approximating rational $r = \frac{a}{b}$ if the desired inequality is to hold. In other words, there are only finitely many (in fact, fewer than d) *different* rationals which lie closer to $s = \frac{c}{d}$ than the square of their denominator.

This suggests the following definition.

Definition:

The real number x is *approximable by rationals to order n* if there is a number K, depending only on x, such that we can find *infinitely many* rationals $\frac{a}{b}$ that satisfy the condition $\left| x - \frac{a}{b} \right| < \frac{K}{b^n}$.

With this definition, any rational number is approximable by (other) rationals to order 1, since the identity $|r - s| = \frac{|ad-bc|}{bd}$ shows that the rational $s = \frac{c}{d}$ is approximated to within $\left(\frac{|ad-bc|}{d} \right) \frac{1}{b}$ by a rational $r = \frac{a}{b}$, and for given c, d we can find infinitely many a, b to satisfy the equation $ad - bc = 1$ (since, with c, d known, this is a single linear equation in the two unknowns a, b). Since d is a natural number, $\frac{1}{d} \leq 1$, and it follows that, if we choose any $K > 1$, there are infinitely many rationals $\frac{c}{d}$ such that $\left| s - \frac{a}{b} \right| < \frac{K}{b}$.

However, the condition $\left| s - \frac{a}{b} \right| < \frac{1}{b^2}$ is possible only if $b < d$, so it follows that s is *not* approximable by rationals to order 2 (and hence to any order higher than 1). Whatever the choice of the constant K (which depends only on s) we have fewer than Kd different choices of $r = \frac{a}{b}$, with a, b relatively prime, such that $\left| \frac{c}{d} - \frac{a}{b} \right| < \frac{K}{b^2}$.

Recall that a real algebraic number a is the root of a polynomial with rational coefficients. We say that the *degree* of a is the *smallest* degree of a polynomial which has a as a real root. From the above, orders of approximation appear to be be connected to the degree of the polynomial that defines a given algebraic number—after all, among algebraic numbers it is the *rationals* that are defined by polynomials of degree 1.

Liouville proved the following remarkable theorem:

A real algebraic number of degree n is not approximable by rationals to any order greater than n.

(The proof makes more significant demands on the reader's background and technical facility than we have done hitherto, and is omitted here. The interested reader can find a version of the proof in *MM*.)

He now had a criterion for determining that certain numbers must be transcendental:

Corollary: *Any real number approximable by rationals to all orders must be transcendental.*

Liouville proceded to define a number that he could prove to be transcendental, based on the above theorem. He considered the sum of an infinite series of negative powers of 10, using only 0 or 1 in the numerator, with the kth use of the digit 1 occurring after $k!$ steps, that is, as $\frac{1}{10^{k!}}$. Thus the first 1 appears in the first decimal place, the second in the second place ($2! = 2$), the third in the sixth place ($3! = 6$), the fourth in the 24th place, and so on. This leads to

Liouville's constant: $L = \frac{1}{10^{1!}} + \frac{1}{10^{2!}} + ... + \frac{1}{10^{n!}} + ...$

The decimal expansion of L starts with $0.1100010000000000000000001000...$ Note that the next 1 will occur at the 120th place after the decimal point, the one after that at the 720th place, the next at the 5040th place, and so on.

The series defining L can easily compared with a geometric one in order to prove that L is a real number. Using the theorem Liouville proved, it is then easy to see that L must be transcendental.

The kth partial sum of the series defining L can be written as $r_k = \frac{p}{10^{k!}}$, where

$$p = 10^{k!-1} + 10^{k!-2} + 10^{k!-6} + 10^{k!-24} + ... + 10^{k!-l!} + ... + 1$$

becomes the numerator when the first $k!$ terms are combined, since all other negative powers of 10 up to $k!$ have numerator 0. All terms of the series are positive, so $r_k < L$. The distance between L and its kth truncation is

$$L - r_k = \frac{1}{10^{(k+1)!}} + \frac{1}{10^{(k+2)!}} + ...$$

How should we compute this infinite sum?

We can estimate the terms on the right by taking out the first term as a common factor and comparing the remaining terms with those of a geometric series:

$$L - r_k = \frac{1}{10^{(k+1)!}}\{1 + \frac{1}{10^{(k+2)}} + \frac{1}{10^{(k+2)(k+3)}} + ...\}$$
$$< \frac{1}{10^{(k+1)!}}\{1 + \frac{1}{10} + \frac{1}{10^2} + ...\}$$
$$= \frac{1}{10^{(k+1)!}}\left(\frac{10}{9}\right)$$

In each case we have replaced each of the exponents in the terms in the final brackets, replacing each of $(k + 2), (k + 2)(k + 3), ...,$ by 1, to obtain the second line, since $10^1 < 10^{k+2}, 10^2 < 10^{(k+2)(k+3)}$, etc. Since $\lim_{k \to \infty}(\frac{1}{10^{(k+1)!}}) = 0$, we conclude that $\lim_{k \to \infty} r_k = L$ is in $\mathbb{R}$.

To show that L is approximable by rationals to all orders, fix a natural number n. For any $k \geq n$ we have found a rational $r_k = \frac{p_k}{q_k}$ where $q_k = \frac{1}{10^{k!}}$ and p_k is the corresponding integer p, as above, such that, with $K = \frac{10}{9}$,

$$\left| L - \frac{p_k}{q_k} \right| < K\frac{1}{10^{(k+1)!}} = K\frac{1}{q_k^{k+1}} < \frac{K}{q_k^n}.$$

For arbitrary choices of n, we always find infinitely many such rationals, namely the sequence $r_{n+1}, r_{n+2}, ..., r_{n+m}, ...$ By the above theorem L is approximable by rationals to all orders, and therefore transcendental.

This proof is straightforward because the number L was *chosen* to make the application of the theorem as simple as possible. Liouville's achievement was to find a workable criterion against which certain numbers could be tested for rapid approximation by rationals.

Proofs that the 'well-understood' numbers π (defined by a geometric relationship, or as the sum of *Leibniz' series* $4(1 - \frac{1}{2} + \frac{1}{5} - \frac{1}{7} + \frac{1}{9} - \frac{1}{11} + ...))$ or e (defined as the sum of $1 + \frac{1}{1!} + \frac{1}{2!} + \frac{1}{3!} + ...$, or as $\lim_{n\to\infty}(1 + \frac{1}{n})^n$), are transcendental, require much more advanced techniques that we cannot pursue here. The transcendence of e was proved by *Charles Hermite* (1822-1901) in 1873, and this was followed in 1882 by a proof of the transcendence of π by *Ferdinand von Lindemann* (1852-1939). He used Hermite's approach to show that e^b is transcendental for any non-zero algebraic number (real or complex). By Euler's identity $e^{i\pi} = -1$, this showed that $i\pi$, and hence π, must be transcendental.

The influential German mathematician David Hilbert famously posed a list of 23 problems at the meeting in Paris in 1900 of the International Congress of Mathematicians. The seventh problem asked whether, given algebraic numbers a, b, with b irrational, the number a^b is always transcendental. In 1934, two mathematicians, *Alexander Gelfond* and *Theodor Schneider*, proved independently that this is true, providing a unified way of solving many problems that had been posed in number theory. Their result shows, for example, that $2^{\sqrt{2}}$ is transcendental.

CHAPTER 8

Axioms for number systems

There are and can be but two ways of investigating and discovering Truth. The one leaps from the senses and particulars to the most general axioms and from these as first principles and their unshaken truth, judges on and discovers medial axioms: and this way is in vogue. The other raises axioms from the senses and particulars, by ascending steadily, step by step, so that at last the most general may be reached; and this way is the true one, but untried.

Sir Francis Bacon, *Novum Organum, Aphorisms I*, 1620

Summary

Having shown how the number systems $\mathbb{Z}$, $\mathbb{Q}$, $\mathbb{R}$ and $\mathbb{C}$ can be constructed by taking the natural numbers as 'given' (as *Kronecker* insisted), we now consider how, in the late 19th century, the *axiomatic* approach, famously pioneered for geometry in Euclid's *Elements*, was applied to the system $\mathbb{N}$ of natural numbers to derive its properties. To explore the reasons behind this shift in perceptions we discuss how the pre-eminence of Euclidean geometry was challenged in the early nineteenth century, in time leading to a re-evaluation of methodology in many areas of mathematics, and setting the scene for current perceptions of the *axiomatic method* throughout the subject.

We then consider the Peano axioms for the natural number system, indicating how the set $\mathbb{N}$, which (with the addition of 0) was the starting point for Dedekind's integers, provides a model for these axioms, while also outlining some differences in the approaches taken by Peano and Dedekind to the same question. This leads on to an axiomatic description of the real number system as a *complete ordered field* containing $\mathbb{Q}$. In the Appendix we show that $\mathbb{C}$ cannot be ordered in a way that is compatible with the algebraic operations.

1. The axiomatic method

In the 1920 edition of his *Introduction to Mathematical Philosophy* (Allen and Unwin, London), the eminent philosopher *Bertrand Russell* (1872-1970) provides a succinct statement of his philosophy of mathematics.

https://doi.org/10.11647/OBP.0236.08

Mathematics is a study which, when we start from its most familiar portions, may be pursued in either of two opposite directions. The more familiar direction is constructive, towards gradually increasing complexity: from integers to fractions, real numbers, complex numbers; from addition and multiplication to differentiation and integration, and on to higher mathematics. The other direction, which is less familiar, proceeds, by analysing, to greater and greater abstractness and logical simplicity; instead of asking what can be defined and deduced from what is assumed to begin with, we ask instead what more general ideas and principles can be found, in terms of which what was our starting-point can be defined or deduced.[1]

In **Chapter 7** we employed, for ease of recognition, modern mathematical terminology in our description of number systems and when discussing how Dedekind and others addressed the extension of number systems in the latter half of the nineteenth century. In doing this, we took the basic number system, the natural numbers, simply as given, without exploring how it might be derived from yet more basic concepts. Our interest now turns to what Bertrand Russell, in the above passage, calls the 'opposite direction', i.e. looking for 'general ideas and principles' from which our earlier starting point (i.e. counting!) might be derived. This search began in earnest only in the late nineteenth century, notably with Dedekind, whose formulation was adapted successfully by Giuseppe Peano to yield *axioms* for $\mathbb{N}$ that eventually gained universal acceptance.

In geometry, what is now called the *axiomatic method* dates back at least to Euclid. His *Elements* begin with a small number of unproven statements, called *postulates* by Euclid, from which all 465 propositions in his thirteen books are then derived by logical deduction and illustrated in geometric constructions.

We might ask how Euclid arrived at his postulates, and whether his concept of the *axiomatic method* is the same as the modern one. The current *Encyclopedia Britannica* has the following definition:

Axiomatic method, *in logic, a procedure by which an entire system (e.g. a science) is generated in accordance with specified rules by logical deduction from certain basic propositions (axioms or postulates), which in turn are constructed from a few terms taken as primitive. These terms and axioms may either be arbitrarily defined and constructed or else be conceived according to a model in which some intuitive warrant for their truth is felt to exist.*

1.1. Axioms in Euclid. The above statement of the axiomatic method in some measure follows the example of Euclid's *Elements*, identifying the 'primitive terms' as well as the 'basic propositions' that will be taken for granted. However, Euclid and his successors, right up to the nineteenth century, would have taken issue with the bold claim in *Encyclopedia Britannica* that the axioms can be *arbitrarily defined and constructed*! For the Greeks,

---

[1]Had Bacon and Russell been contemporaries, they might had much to say to each other!

and for many centuries after them, the notion of an *intuitive warrant for their truth* was deemed to be an absolutely central requirement for any system of axioms. In geometry this relied on a notion of geometrical truth: properties of lines, angles, polygons and circles in the geometry described by Euclid, as well as their three-dimensional counterparts in solid geometry, were understood as representations of observed spatial reality, albeit in an idealised form. The truth of the underlying axioms was expected to reflect *self-evident* aspects of this observed reality.

The most significant philosophical shift leading to the modern understanding of axioms as described in the above definition is that it is the axioms themselves that should *determine* the structure of the collection of objects under discussion. In this approach the 'primitive terms' stated at the outset have no *intrinsic* meaning independently of the axioms. This viewpoint emerged gradually during the nineteenth century, notably in alternatives to Euclidean geometry produced by Gauss, Bolyai and Lobachevsky, but also in Hamilton's work on complex numbers that we outlined in **Chapter 4**.

Following a distinction first made by Aristotle, Euclid states five 'common notions' and five 'postulates'. The former are assumptions that would apply to any quantitative science, such as: 'When equals are added to equals, the results will be equal', and Euclid's now famous dictum: 'The whole is greater than the part'. The postulates are more specific: they are the unproven assumptions he accepts for his geometry.

Skipping lightly over the common notions, we focus on Euclid's five *postulates*. In order to state these, his text introduces basic notions such as 'point, 'line' and 'circle' (what the *Britannica* definition calls 'primitive' terms). The first four basic notions famously read:

1. A *point* is that which has no part.

2. A *line* is breadthless length.

3. The extremities of a line are points.

4. A *straight line* is a line which lies evenly with the points on itself.

In all, Euclid lists 23 such definitions, that of the *circle* being the fifteenth: A *circle* is a plane figure contained by one line such that all the straight lines falling upon it from one point among those lying within the figure are equal to one another. (The point in question is the centre of the circle.)

Historians have debated these rather odd definitions.[2] Note that 'points' are defined both in 1. and 3. In 2., the terms 'length' and 'breadth' are assumed (implicitly) as known, and 3. suggests that the 'line' is finite, that is, a 'line segment' or an 'arc'. Definition 4. seeks to identify 'straight' lines, but its meaning is at best opaque.

---

[2]A gentle introduction to this debate can be found at mathshistory.st-andrews.ac.uk/Hist Topics/Euclid_definitions.html, for example. For a detailed account see **[21]**.

It has been argued that the list of definitions may be a later addition to Euclid's work, and not by himself, since he never refers back to them in the *Elements*, and later introduces terms (such as *magnitude*) that he has not defined at all. If that is the case, then Euclid's approach was truly modern!

In his five postulates, Euclid identified specific assumptions taken as *self-evident* (not requiring proof) that he needed to provide the basis of his geometry. Everything else he asserted in the *Elements* would be deduced logically from these five postulates. Euclid's *deductive approach* has served as the paradigm for the textbook transmission of mathematical knowledge for more than two millennia.

The first four of Euclid's five postulates for geometry may indeed seem self-evident: the first three simply assert that it is possible to

(i) draw a straight line from any point to any point,

(ii) produce a finite straight line continuously in a straight line,

(iii) describe a circle with any centre and diameter.

These postulates really describe Euclid's perception of space: it is *continuous* (there are no gaps) and *not limited.*

The fourth postulate states that

(iv) all right angles are equal to one another.

Euclid's definition of *right angle* is given in terms of two lines cutting each other:

When a straight line standing on a straight line makes the adjacent angles equal to one another, each of the equal angles is right. Postulate (iv) says that angles produced in this manner must have the same magnitude, wherever they are in space. Thus he assumes that space is *homogeneous,* i.e. figures retain their shape wherever they are placed.

The fifth postulate—generally known as the *parallel postulate*—is key to much of the *Elements*. In formulating the postulate, Euclid is careful not to talk about what we call parallel lines although he had earlier defined these. Instead:

(v) *'If a line segment intersects two straight lines forming two interior angles on the same side that sum to less than two right angles, then the two lines, if extended indefinitely, meet on that side on which the angles sum to less than two right angles.'*

Many elementary texts are less careful than Euclid: they may 'define' parallel lines as 'lines that never meet', or that 'meet at infinity'. Both phrases beg obvious questions. In the better school textbooks Euclid's formulation of the parallel postulate is often replaced by a logically equivalent version. In this form it is usually credited to the eighteenth-century Scottish mathematician *John Playfair,* but it was stated in similar terms by Proclus in the fifth century:

Given a line l and a point P not on l, in the plane containing both P and l there is exactly one line through P that does not intersect l.

This unique line is then called the line *parallel* to *l* through *P*.

1.2. The impact of non-Euclidean geometry. Euclid needed all five postulates to prove the 465 propositions he formulated in the *Elements*. However, he avoided using the fifth postulate for as long as possible (Propositions 1-28 do not use it), which may suggest that he could also have regarded it as less obviously 'self-evident' than the other four. For over 2000 years, natural philosophers and mathematicians alike accepted Euclidean geometry as the proper mathematical description of the space we live in. This belief required the self-evident nature of the postulates to remain firmly grounded in visual perception. The fifth postulate, however, makes implicit assumptions about the nature of space: how would we know that two 'parallel' lines might not meet in some far-off region of space?

Some of Euclid's successors in antiquity (such as Ptolemy) and early commentators (e.g. Proclus) thought that the parallel postulate did not appear to be fully self-evident. They therefore tried to deduce it from the other four postulates, which they did regard as indisputably true.

The search for a proof of the fifth postulate resurfaced repeatedly over two millennia, throughout the transmission of the *Elements* to Arab lands and Europe. Prominent European mathematicians, John Wallis and *Adrien-Marie Legendre* (1752-1833) among them, set out to prove the postulate, but it was the Italian Jesuit priest *Giovanni Saccheri* (1667-1733) who took a decisive step, seeking to prove the postulate by showing that alternative hypotheses would lead to contradictions. He showed that the postulate was logically equivalent to the claim that the angle-sum of a triangle equals two right angles, and tried to prove, using only the other four postulates, that this angle-sum could be neither more nor less than this. But, try as he might, he was unable to find arguments that would contradict the 'hypothesis of the acute angle': that the angle-sum was *less* than two right angles.[3]

Although Saccheri's work suggested a new approach to the problem, it took a further century for it to be resolved. In the 1820s, the Russian *Nikolai Lobachevsky* (1793-1856) and the Hungarian *Janos Bolyai* (1802-1860) independently demonstrated the possibility of *non-Euclidean* geometry, in which the first four postulates remain valid, but Euclid's fifth postulate is replaced by the hypothesis of the acute angle. In terms of Playfair's formulation, in their *hyperbolic geometry* Euclid's fifth postulate was replaced by the statement:

---

[3]In spherical geometry *great circles* (geodesics) play the role that 'straight lines' play in Euclidean geometry. The angle-sum of a triangle on the surface of a sphere is greater than two right angles. But geodesics are finite, so cannot be extended indefinitely (without meeting themselves). This led Saccheri to reject the hypothesis that the angle-sum in a triangle can exceed two right angles.

Given a line l and point P not on l, in the plane containing both P and l there are at least two distinct lines through P that do not intersect l.

While they were the first to publish their results (Lobachevsky in 1829 and Bolyai in 1832-3, although he had completed a full draft of his work by 1823), they were not the only discoverers of the new geometry. The notebooks and correspondence of Carl Friedrich Gauss, preserved after his death, show conclusively that he had studied the fifth postulate extensively for decades, was convinced by 1816 that it could not be proved, and had developed the bulk of the theory of hyperbolic geometry by the following year.[4]

Gauss never published his own results. In a letter to his friend, the prominent mathematician, physicist and geodesist *Friedrich Wilhelm Bessel* (1784-1846), he states the main reason for his reluctance: *'da ich has Geschrei der Boeoter scheue'* ('since I dread the shouts of the Boethians'). This is a reference to the—arguably malign—influence on European mathematics of the mediocre text *De Institutione Arithmetica* by the Roman scholar Boethius, which (as we saw in **Chapter 2**) had been the main source of Greek mathematics available in Western Europe until the late Middle Ages.

Gauss' assessment of his contemporaries' readiness for such a radically new theory may well have been accurate. One important reason is that the possibility of any geometry other than that of Euclid flatly contradicts the claim by the influential philosopher *Immanuel Kant* (1724-1804) that our concept of space is given *a priori* as that of Euclidean space. This Latin phrase means *'from the earlier'*. Thus, *a priori* reasoning reaches conclusions that arise necessarily from first principles (premisses), rather than relying on observation or experiment. The term was made popular by Kant in his *Critique of Pure Reason*, published in 1781.

In his extensive correspondence with Bessel, Gauss is very clear that his perception of Euclidean space differs radically from that of Kant. Bessel, in turn, confirms that he shares Gauss' viewpoint. In a letter to Gauss on 10 February 1829, he agrees that *'our geometry is incomplete'*, and pleads with Gauss to ignore the 'Boethians'. Gauss responding on 9 April 1830, ignores this plea, but expresses his *'innermost conviction'* that

'geometry stands in a quite different position from pure arithmetic with regard to our a priori knowledge: this [knowledge] *rests on our complete conviction of its necessity (thus also of its absolute truth), which the latter possesses. We must admit in all humility that,* [even] *if number is a pure product of our mind, space has an external reality, for which we cannot prescribe its laws a priori.'*

---

[4]The near-simultaneous discovery of non-Euclidean geometry by several practitioners adds weight to an oft-repeated saying attributed to Gauss – here taken from E.T. Bell's *The Development of Mathematics* (1945): *Mathematical discoveries, like springtime violets in the woods, have their season which no human can hasten or retard.*

In stark contrast to Gauss' then unpublished comments, many mathematicians regarded Bolyai and Lobachevsky's publications as controversial in the 1830s, and they were by no means universally accepted as groundbreaking achievements. Significant changes in these perceptions had to wait for the decade after Gauss' death (in 1855), when publication of his private papers revealed the great man's interest in and strong views on this subject. At the same time, substantial new work by his former student *Bernhard Riemann* (1826-1866) in 1854 and in 1868 by the Italian geometer *Eugenio Beltrami* (1835-1900)—the first to provide a model of hyperbolic geometry—finally led to general acceptance that non-Euclidean geometries were just as consistent as Euclid's geometry, and should be accorded equal status.

These developments ushered in the modern perspective on the role of axioms in different areas of mathematics. For example, it was now clear that different 'geometries' with quite different characteristics can be explored, depending on which axioms one chooses to adopt. Nor was all this of importance only in 'pure' mathematics, and devoid of physical meaning, as even Riemann had supposed. By the early twentieth century, *Albert Einstein* (1879-1953) and *Hermann Minkowski* (1864-1909) had shown that hyperbolic geometry provided an appropriate mathematical model for relativistic space-time. In the same period, *David Hilbert* (1862-1943) updated Euclid's postulates, requiring 20 axioms for Euclidean geometry, most of which were needed to fill logical gaps that Euclid had left by relying on visual perception – for example, the intuitive notion of an object 'lying between' two others had to be given axiomatic form.

Beyond geometry, the renewed focus on axiomatics meant that mathematicans sought to formulate axiom systems to make explicit the assumptions on which their particular areas of interest, as well as the foundations of the subject, can be based. In shifting the emphasis from discussion of the 'nature' of specific objects to *'more general ideas and principles'* (as Russell puts it), it is the *rules for interaction* between mathematical objects that become paramount. The axioms that specify these rules often find application in a variety of different situations: although they may have originated in the analysis of a particular set of problems, similar techniques can find application in an apparently unrelated field. Complex numbers, for example, allow us to calculate freely with the 'imaginary unit' i according to Hamilton's rules, even though we can envisage no concrete physical representation for it (as we might do for $\sqrt{2}$ as the length of the diagonal of a unit square). Yet i is central to the analysis of sinusoidal waves by electrical engineers (who insist on calling it j) where it measures the 'imaginary' flow of electricity through various bodies.

The axiomatic method, based on set theory as the keystone of modern mathematics, has opened up a vast range of new areas for exploration since the nineteenth century, most of them far beyond the scope of this book. For the most part, the question of what the objects being considered 'are' has been replaced by asking what they *do,* that is, how they interact according to the underlying axioms of their particular field. But this does not imply that every new theorem is painstakingly traced back to these axioms – mathematics is a cumulative undertaking, and what has previously been confirmed as correct can be used to justify new assertions.

In practice, most mathematicians choose to work on the implicit assumption that the currently accepted system of axioms for set theory (developed in the first decades of the twentieth century) can be shown to underpin their researches.[5] Very few, however, would undertake the arduous task of checking complex deductions by direct reference to theseaxioms! There are well-known examples of proofs that—written out in full—would run to hundreds, sometimes thousands, of pages of logical deduction and computation. In most branches of modern mathematics there are a good many results that have only been checked in full detail by a small group of qualified experts. In the end, the 'truth' of a mathematical statement comes down to its (preferably unanimous) acceptance, after a suitably exhaustive analysis and confirmation, by the mathematical community of the time. As *Davis* and *Hersh* put it forcefully in their book *The Mathematical Experience*: in practice, the validity of a proof in mathematics is established by a *'consensus of the qualified'.* (Of course, even the 'qualified' are not infallible.)

2. The Peano axioms

Let us be modest and remain with our naive concept of a set, as in Cantor's formulation quoted in **Chapter 7**, Footnote 1. The axioms described below for *arithmetic* with the natural numbers are essentially those devised by the Italian mathematician Giuseppe Peano, whose work was deeply influenced by an analysis of number systems published by Dedekind in 1888 as *Was sind und was sollen die Zahlen?*[9]. This translates literally as 'What are numbers and what should they be?'[6] In his pamphlet Dedekind spells out his philosophy, beginning his Preface with the sentence: *'In Science, what is provable should not be believed without proof.'* He argues that, although obvious, this demand has by no means been reached in the foundation of the simplest science, namely *'the part of logic dealing with the study of numbers'.* Designating arithmetic (and hence algebra and analysis) as a part of logic implies that he regards the number concept as entirely independent of conceptions of space or time, but rather as an immediate consequence of *'pure laws of thought'.* His answer to the question raised in the title of his work is

---

[5]These are known as the *Zermelo-Fraenkel axioms, (ZF)*, after their originators *Ernst Zermelo* (1871-1953) and *Abraham Fraenkel* (1891-1965). See Chapter 10 and *MM* for a brief discussion.

[6]An English translation in 1901 by Wooster Woodruff Beman (1850-1922) renders the title as 'The Meaning of Numbers'.

that 'numbers are free creations of the human mind, they serve as a means of understanding the diversity of objects ['Dinge'] more easily and sharply'. Only by having developed the science of numbers by purely logical means, arriving at the 'continuous number-domain' (the real numbers), can we examine our perceptions of space and time more precisely, he claims.

Rather than focus on the details of Dedekind's pamphlet here, we turn to the version of these ideas as presented by Peano, before giving a brief comparison of the two approaches.

Peano acknowledged the profound impact of Dedekind's work on his own formulation. He published this in 1889 in a pamphlet entitled *Arithmetices principia, nova methodo exposita* ('Principles of arithmetic, expounded by a new method'). The axioms are sometimes called the *Dedekind-Peano axioms*.

Peano used the concept of *successor* as his starting point. Roughly speaking, the system of five axioms he postulated enables us to show that all familiar arithmetical properties of the natural numbers can be deduced provided we can *start* with an element we call 1 (as the Greeks did with the unit), and that we can apply *induction*.

Following Peano, the three 'primitive terms' we begin with are therefore: an *abstractly* given set we will call $\mathbb{N}$, together with a 'distinguished element' that we denote by 1 and a *successor operation* S between members of $\mathbb{N}$.

The five *Peano axioms* are:[7]

1. 1 is a member of $\mathbb{N}$

2. every n in $\mathbb{N}$ has a successor $S(n)$ in $\mathbb{N}$

3. if $S(m) = S(n)$ then $m = n$

4. 1 is *not* the successor of any element of $\mathbb{N}$

5. (*Induction*) if A is a subset of $\mathbb{N}$ with $1 \in A$ and such that $n \in A$ implies $S(n) \in A$, then $A = \mathbb{N}$.

Axiom 2 tells us that taking successors keeps us within the set, axiom 3 that an element cannot be the successor of two different elements and axiom 4 means that 'nothing comes before' the distinguished element we call 1.

Note also that any element of $\mathbb{N}$ other than 1 must be the successor of some element of $\mathbb{N}$. To deduce this from Axiom 5, let S be the set consisting of 1 and all successors; that is, S contains 1 together with every n in $\mathbb{N}$ which is the successor of some element of $\mathbb{N}$. Then 1 belongs to S by definition,

---

[7]Some authors prefer to state our axioms 1. and 4. below as the single statement: 'there is exactly one member of $\mathbb{N}$ which is not the successor of any other member of $\mathbb{N}$'. Peano himself also included four further axioms to clarify the use of *equality* ($=$), stipulating that the set he is describing should be *closed* under this relation and that equality is an *equivalence relation* – see Footnote 4 in **Chapter 7** for this terminology.

while, if $m \in S \subseteq \mathbb{N}$, then its successor $S(m)$ must lie in S, since S contains *all* the successors of members of $\mathbb{N}$. By Axiom 5 this means that $S = \mathbb{N}$.

Consider the sequence

$$1, S(1), S(S(1)), S(S(S(1))), ...$$

where repeated use of S just means that we take the 'successor of the successor' and so on. We can give these successive elements 'abbreviated names'. For example, we could decide to denote them by $2 = S(1)$, $3 = S(2) = S(S(1))$, and so on. Then, after (long-scale!) octillion repetitions we would arrive at what we can denote by $10^{48} + 1$. The process does not stop after any finite number of repetitions.

We will now *represent* the set $\mathbb{N}$ by the sequence $\{1, 2, 3, ..., n, ...\}$. But it must be stressed that, in giving 'names' to the 'distinguished element' 1 and to its various successors we do *not attach a particular meaning* to the symbols we use. The 'meaning' we attach to a symbol is *determined by* the position it occupies. As in Dedekind's derivation of the structure of the sets of integers and rationals (and in Hamilton's definition of the complex numbers) it is the *structure* that gives meaning to the various members of the set. For $\mathbb{N}$ this structure is determined by the Peano axioms alone.

We can outline the differences between the approaches of Peano and Dedekind more easily by following [6] in using Peano's axioms as our point of reference. Dedekind's starting point is the concept of *order*. To do this, he first considers *mappings* between abstract sets, then specialises to one-to-one mappings from a set to itself—that is, ϕ maps S to itself—and distinct elements have distinct images.

If a set N has such a mapping ϕ and there is an element t in N such that ϕ does *not* map any element of N to t then Dedekind calls N a *chain* [Kette]. We can see how these requirements mirror Peano's first four axioms. The fifth axiom (induction) now becomes the statement that N *is* the chain generated by the (consecutive) images of this distinguished element t. He shows that such a set is a particular kind of infinite set, which he calls *simply infinite* (in **Chapter 9** we will consider such sets, nowadays called *countably infinite*). Dedekind also showed that any infinite set has a simply infinite subset. However, his 'proof' that infinite sets *exist* was fiercely criticised at the time, especially by the logician *Gottlob Frege* (1848-1925).

Nonetheless, it should be reassuring to know that Dedekind proved that any two *models* of the above axioms – that is, well-defined sets that give meaning to the primitive terms and satisfy the stated axioms – will have the same mathematical structure (the formal term is that they are *isomorphic*). We could have listed the elements of the set $\mathbb{N}$ in a variety of ways. For example, we might take 1 followed in succession by all positive integral powers of 10, with $S(10^k)$ defining 10^{k+1}, yielding $\{1, 10, 100, 1000, ...\}$.

As long as the axioms are satisfied, this structure will be isomorphic to our $\{1, 2, 3, ..., n, ...\}$.

Peano defined the *operations of arithmetic*, namely addition and multiplication, as follows:

Addition is defined *recursively* for elements m, n in $\mathbb{N}$, first by adding one, i.e. $n + 1 = S(n)$, and then, for any m, n, setting

$$n + S(m) = S(n + m).$$

The first of these definitions just gives the successor of n the name $n + 1$.

To continue, we proceed in steps, using the 'names' of successors $2 = S(1), 3 = S(2), 4 = S(3)$ etc.: take $m = 1, n = 2$, in the general definition, so $2 + S(1) = S(2 + 1) = S(3) = 4$. We have shown that $2 + 2 = 4$, and can continue in this fashion (see also [6] , p.255).

Multiplication can be defined similarly:

(i) $n \times 1 = n$ for all n in $\mathbb{N}$,

(ii) for n, m in $\mathbb{N}$, $n \times S(m) = n \times m + n$.

(It is again instructive – if lengthy – to test this definition with various examples.)

The (strict) *order* relation $<$ on $\mathbb{N}$ is then defined as before: $m < n$ if there exists k in $\mathbb{N}$ such that $m + k = n$. Again, $m \leq n$ will mean that either $m < n$ or $m = n$.

From these definitions one can derive the *laws of arithmetic* and *order properties* of the set $\mathbb{N}$. These steps will be omitted here – as in **Chapter 7**, detailed proofs can be found in [30].

As we observed in **Chapter 7,** starting induction with 0 instead of 1 creates no new difficulties, and we found it useful to treat 0 as a 'natural number', i.e. to begin with the set $\mathbb{N}_0 = \{0, 1, 2, ..., n, ...\}$ in our discussion of Dedekind's representations of $\mathbb{Z}$ and $\mathbb{Q}$. Similarly, we could have used 0 to replace 1 in the Peano axioms. This version of *Peano arithmetic* (starting with a set we call $\mathbb{N}_0$, a distinguished element 0, and a successor operation S as before) is essentially the same as what we outlined above.

Another argument for including 0 is based on the desire to 'start from scratch'; that is, to build up a number system using only basic *operations on sets*. The Hungarian-American mathematician *John von Neumann* (1903-1957) showed how, starting with the empty set $\varnothing$, the set $\mathbb{N}_0$ has a simple representation as an infinite collection of sets, each of which has finitely many elements.

Here is von Neumann's construction of $\mathbb{N}_0$ by starting with the empty set:

$\varnothing$—a set with no elements—is represented by the symbol 0,

Figure 38. John von Neumann, from a period while at Los Alamos National Laboratory, from *Los Alamos: Beginning of an era, 1943-1945,* Los Alamos Scientific Laboratory, 1986[8]

$\{\varnothing\}$—a singleton set—is represented by the symbol 1,

$\{\ \varnothing, \{\varnothing\}\}$—a set with two elements—is represented by the symbol 2,

$\{\varnothing, \{\varnothing\}, \{\varnothing, \{\varnothing\}\}\}$ is represented by 3, and so forth.

Together, these symbols are a representation of the set of natural numbers. Having started with the 'distinguished' element 0, (as given by $\varnothing$), the successor operation is given by

$$S(n) = n \cup \{n\}.$$

This makes sense, since each 'number' n is a set (as listed above for $0, 1, 2, 3$) and to obtain the successor of n we simply take the set A that represents n and its union with the set $\{A\}$ whose only element is A. It can be checked that von Neumann's construction satisfies the amended version of Peano's axioms 1-5, with 0 playing the role of 1. The definitions of addition and multiplication stay the same as stated above.

The *number system* $(\mathbb{N}_0, +, \times, <)$ then provides a *model* of the natural numbers (including 0 this time), which in turn form the basis of arithmetic. This is why mathematics is sometimes jokingly described as *'the theory of the empty set'*!

---

[8]https://commons.wikimedia.org/wiki/File:JohnvonNeumann-LosAlamos.jpg

3. AXIOMS FOR THE REAL NUMBER SYSTEM

3. Axioms for the real number system

Having seen how the properties of the number system $\mathbb{N}$ can be derived from the Peano axioms, and how each of the systems $\mathbb{Z}, \mathbb{Q}$ and $\mathbb{R}$, and their arithmetical and order properties, can in turn be derived from its predecessor, we now list the fundamental properties that the system $\mathbb{R}$ should satisfy.

The axioms we need to check for any model of the real number system $\mathbb{R}$ are divided into three groups:

1. *Algebraic axioms*

2. *Order axioms*

3. *Completeness axiom*

The first two groups were essentially given in **Chapter 7**, but we repeat them more fully here. To do so effectively, without presupposing that we can call the object we are defining a 'number system', we need to give abstract definitions of the algebraic operations on and order relations between the elements of an (initially) unspecified *set*.

We introduce terminology to describe addition and multiplication abstractly: addition and multiplication of two numbers associate a pair of elements of a set $\mathbb{S}$ with an element of $\mathbb{S}$ (formally, they are functions from $\mathbb{S} \times \mathbb{S}$ to $\mathbb{S}$); we write $(a, b) \to a + b$ and $(a, b) \to a \times b$. We call these *binary operations* and, for our number systems, the laws of arithmetic are described by the axioms these operations should obey.

The essential point remains that we are less concerned with defining the *objects* of our number system, but focus our attention instead on the *relations* between them – in other words, we describe how they *interact*. Provided we spell out the requirements for the binary operations $+, \times$, we can apply them to any (unspecified) non-empty set $\mathbb{S}$.

Similarly, for any set $\mathbb{S}$ we can identify a relation, i.e. a collection O of pairs (a, b) with a, b in $\mathbb{S}$, for which we write $a \leq b$ if and only if $(a, b) \in O$. As indicated earlier, this defines a *linear order* (also called a *weak total* order) if it satisfies the following three requirements:

1. *Antisymmetry: if $a \leq b$ and $b \leq a$ then $a = b$,*

2. *Transitivity: if $a \leq b$ and $b \leq c$ then $a \leq c$,*

3. *Totality: For any a, b in $\mathbb{S}$, either $a \leq b$ or $b \leq a$ (or both).*

The third property says that any two elements a, b of S can be compared: at least one of the pairs (a, b) and (b, a) belongs to O. Note that this includes the possibility that both belong to O, but, in that case, the first property shows that we equate a and b, and write $a = b$.

The concepts of *upper bound* and *least upper bound*, which we encountered in **Chapter 7**, clearly make sense for non-empty subsets of any linearly ordered set.

3.1. Axioms for a complete ordered field. The definitions below will therefore make no direct reference to the set of $\mathbb{R}$ real numbers, but will instead list three groups of axioms. A system consisting of a set together with two binary operations which satisfy the first group of axioms listed below as Definition 1(a) - the 'algebraic' axioms – is called a *field*. If these operations are compatible with an ordering as described in Definition 1(b) – the 'order' axioms – then it becomes an *ordered field*. Finally, if the axiom stated as Definition 1(c)—completeness—is satisfied in addition, we have a *complete ordered field*. We have a choice of several equivalent statements to describe completeness. We will take the existence of a supremum for each set that is bounded above as our axiom.

*Definition 1(a): A **field** is a set $\mathbb{S}$, together with two binary operations, i.e. functions from $\mathbb{S} \times \mathbb{S}$ to $\mathbb{S}$. They are: addition, denoted by $+$, and multiplication, denoted by $\times$, so that, applied to (a, b), addition gives $a+b$ and multiplication gives $a \times b$ as members of $\mathbb{S}$. These operations have the following properties:*

Algebraic axioms

(i) for all $a, b \in \mathbb{S}$, $b + a = a + b$ and $b \times a = a \times b$ (commutative laws),

(ii) for all $a, b, c \in \mathbb{S}$, $a + (b+c) = (a+b) + c$ and $a \times (b \times c) = (a \times b) \times c$ (associative laws),

(iii) for all $a, b, c \in \mathbb{S}$, $a \times (b+c) = a \times b + a \times c$ (distributive law),

(iv) there are elements of $\mathbb{S}$ denoted by $0, 1$ respectively, such that $0 \neq 1$ and $a + 0 = a$, $a \times 1 = a$ for all $a \in \mathbb{S}$ (existence of neutral elements),

(v) for each $a \in \mathbb{S}$ there exists an element of $\mathbb{S}$ denoted by $-a$, such that $a + (-a) = 0$; for each $a \neq 0$ in $\mathbb{S}$ there exists an element of $\mathbb{S}$, denoted by a^{-1}, such that $a \times a^{-1} = 1$ (existence of inverses).

With the algebraic axioms in Definition 1(a) we define the *difference* $a - b$ as $a + (-b)$, and the *quotient* (usually written as $\frac{a}{b}$) as $a \times b^{-1}$ whenever $b \neq 0$. Both of these definitions make sense because of axiom *(v)* in Definition 1(a).

Definition 1(b):

*A relation $<$ on a set $\mathbb{S}$ is a **strict total order** if the following axioms hold:*

Order axioms:

(i) (Anti-reflexivity) $a < a$ is not true for any a in S,

(ii) (Trichotomy) For all a, b in $\mathbb{S}$, exactly one of $a < b$, $a = b$, $b < a$, is true,

(iii) (Transitivity) For $a, b, c \in \mathbb{S}$, if $a < b$ and $b < c$, then $a < c$.

The field $(\mathbb{S}, +, \times)$ *– defined in 1(a) – is an **ordered field** if it has a strict total order* $<$ *satisfying the conditions:*

(iv) if $a < b$ *and* $c \in S$ *then* $a + c < b + c$,

(v) if $a < b$ *and* $0 < c$ *then* $a \times c < b \times c$.

Axioms *(iv)* and *(v)* of Definition 1(b) ensure that the ordering $<$ is *compatible* with the algebraic operations $+$, $\times$. Together, *(a)* and *(b)* in Definition 1 suffice to describe the axioms for the number system $\mathbb{Q}$ as an example of an ordered field.

To reconcile the order axioms in Definition 1(b) with the concept of a *linear* ordering, we need only define $a \leq b$ to mean that either $a < b$ or $a = b$. Then $\leq$ is a linear ordering, as described above: the required properties (1), (2), (3) we listed earlier follow immediately. We define $a > b$ to mean that $b < a$, similarly $a \geq b$ means $b \leq a$. Finally, $a \in \mathbb{S}$ is called *positive* if $a > 0$ and *negative if* $a < 0$.

Moreover, b in $\mathbb{S}$ is an *upper bound* for the subset E of **S** *if* $a \leq b$ for all a in E, and we say that E is *bounded above* by b. Finally, c is the *least upper bound of* E, written as $c = \sup E$, if $c \leq b$ for every upper bound of E.

This terminology allows us to state our final axiom, describing *completeness* as follows:[9]

Definition 1(c): An ordered field $(\mathbb{S}, +, \times, <)$ *is **complete** if the following axiom holds: every non-empty subset of* $\mathbb{S}$ *that is bounded above in* $\mathbb{S}$ *has a least upper bound in* $\mathbb{S}$.

It is proved in *MM* that any complete ordered field $\mathbb{S}$ containing $\mathbb{Q}$ will inherit the Archimedean property from $\mathbb{Q}$: *given positive elements* a, b *of* $\mathbb{S}$, *there is a natural number* n *such that* $na > b$.

3.2. Embedding $\mathbb{Q}$ **in** $\mathbb{R}$. In particular, if a set $\mathbb{S}$ satisfying *(a)-(c)* in Definition 1 is to be a number system that extends the ordered field $\mathbb{Q}$, the algebraic operations and order relation for $\mathbb{Q}$ should be compatible with those for $\mathbb{S}$. If we can embed $\mathbb{Q}$ as a *subset* of $\mathbb{S}$ we will wish, finally, to call this set

---

[9]Proofs of the equivalence of the following five versions of this axiom for **R** can be found in *MM*:

(a) *Bounded non-decreasing sequences property:* any non-decreasing sequence of real numbers, bounded above in $\mathbb{R}$, will always have a limit in $\mathbb{R}$.

(b) *Nested sequences property:* any nested sequence of closed intervals in $\mathbb{R}$ whose lengths decrease to 0 defines a unique element of $\mathbb{R}$.

(c) *Least upper bound property:* any non-empty subset B of $\mathbb{R}$ that has an upper bound in $\mathbb{R}$ has a least upper bound in $\mathbb{R}$.

(d) *Cauchy Criterion:* every Cauchy sequence in $\mathbb{R}$ converges to a real number.

(e) *Bolzano-Weierstrass property:* every bounded sequence of real numbers has at least one subsequence that converges in $\mathbb{R}$.

the *real numbers* and denote it by $\mathbb{R}$. From the perspective of *constructing a model* for $\mathbb{R}$, *Definition 1(a)-(c)* sets out requirements the model must satisfy.

Hence, when we start with $\mathbb{Q}$ and construct a model for $\mathbb{R}$, we require two additional steps to ensure that everything is consistent:

(i) When we use the operations $(r, s) \to r + s$ and $(r, s) \to r \times s$ with *rational* r, s, the outcome should be the same as we had for addition and multiplication in $\mathbb{Q}$. Moreover, $0 < r$ for rational r should mean the same as in Definition 1(b). In other words, the operations should remain the same, regardless of whether we treat r, s as rationals or as real numbers. Formally, this means that we have to construct a function ϕ from $\mathbb{Q}$ to $\mathbb{R}$ such that $r \neq s$ implies $\phi(r) \neq \phi(s)$ and the image $\phi(r + s)$ of their sum equals $\phi(r) + \phi(s)$, where the second $+$ is interpreted as addition in $\mathbb{R}$. Similarly, $\phi(r \times s) = \phi(r) \times \phi(s)$, where $\times$ denotes on the left denotes multiplication in $\mathbb{Q}$ while on the right it denotes multiplication in $\mathbb{R}$. Moreover, the image under ϕ each positive rational r must satisfy $0 < \phi(r)$. This ensures that we can treat $\mathbb{Q}$ as a subset of $\mathbb{R}$.

(ii) The set $\mathbb{R}$ satisfying all the above conditions should be *unique* (else we could not talk about *the* real numbers).

Thus all the different constructions of *models* for $\mathbb{R}$ should 'have the same form', i.e. be *isomorphic*. In other words, it should be possible to put any two of them in a one-one correspondence in such a way that the operations of addition and multiplication, as well as the ordering, correspond correctly. This is the case for Dedekind's and Cantor's models (see [10]).

4. Appendix: arithmetic and order in $\mathbb{C}$

Finally, let us justify a claim made in Section 1 of **Chapter 7**: that the complex number system $\mathbb{C}$ *cannot* be (totally) ordered in a way that is compatible with addition and multiplication. As the laws of arithmetic listed above hold in $\mathbb{R}$, Hamilton's definitions for addition and multiplication will carry them over to $\mathbb{C}$. In $\mathbb{C}$, the pair $(0, 0)$ serves as the neutral element 0 for addition ($0 + z = z$ for any z in $\mathbb{C}$) and any z in $\mathbb{C}$ has an additive inverse $-z$ that satisfies $z + (-z) = 0$. The pair $(1, 0)$ defines the neutral element 1 for multiplication in $\mathbb{C}$.

If we could define a strict total order $\prec$ on $\mathbb{C}$ that is compatible with addition and multiplication, statements (i)-(iv) below would hold for $(+, \times, \prec)$:

(i) *totality*: any two complex numbers can be compared using $\prec$,

(ii) *trichotomy*: given any z in $\mathbb{C}$, exactly one of $0 \prec z$, $z = 0$, $z \prec 0$ is true,

(iii) *Compatibility with* $+$: given any three complex numbers z_1, z_2, z_3, if $z_1 \prec z_2$ then $z_1 + z_3 \prec z_2 + z_3$,

(iv) *Compatibility with* $\times$: given any three complex numbers z_1, z_2, z_3, if $0 \prec z_3$, then $z_1 z_3 \prec z_2 z_3$.

Apply (i)-(iv) with $z = i$, the 'imaginary unit' $(0, 1)$. Since $i \neq 0$, (ii) ensures that either $0 \prec i$ or $i \prec 0$ (not both).

If $0 \prec i$ we have $0 \prec i \times i = -1$ by (iv). Using (iv) again, $0 \prec -1$ shows that $0 \prec (-1) \times (-1) = 1$. But (iii) yields $1 = 0 + 1 \prec (-1) + 1 = 0$. We have arrived at $1 \prec 0 \prec 1$, which is a contradiction for a strict order.

If $i \prec 0$, (iii) implies $0 = i + (-i) \prec 0 + (-i) = (-i)$, and by (iv) again, $0 \prec (-i) \times (-i) = -1$. As in the first case this implies the contradiction $1 \prec 0 \prec 1$. Since both possibilities lead to a contradiction, the order $\prec$ cannot be compatible with addition and multiplication.

CHAPTER 9

Counting beyond the finite

Let every student of nature take this as his rule, that whatever the mind seizes upon with particular satisfaction is to be held in suspicion.

Sir Francis Bacon, *Novum Organum*, 1620

Summary

In this chapter we begin with Georg Cantor's work on the continuum, which reflects the abstract approach he and Richard Dedekind had shared in their models for the real number system. From the 1870s onward, their work was to have a profound influence on the development of mathematics. Within a decade, further investigation into the nature of the continuum would lead Cantor to focus on the nature of *infinite sets*, which sparked deep philosophical disagreements between leading groups of mathematicians about the nature of their subject. This would culminate in a profound conceptual revolution in prevailing views of the nature of mathematical truth.

Cantor's perception of the continuum was to lead him to explore a general notion of 'size' for sets, prompted by 'different kinds of infinity' apparently represented by $\mathbb{Q}$ and $\mathbb{R}$. His investigation of trigonometric series, on the other hand, stimulated his development of the far-reaching concepts of *transfinite* ordinal and cardinal numbers as an abstract method of continuing the process of 'counting' beyond finite sets. His groundbreaking papers in both these areas laid the groundwork for an entirely new *theory of sets*, providing a basis for the whole of mathematics, while at the same time foreshadowing troubling *paradoxes* that would come to plague this new theory.

1. Cantor's continuum

Georg Cantor's initial motivation for investigating the continuum had nothing to do with concerns about the teaching of Calculus to students. Instead, it was his analysis of Fourier series that led him to his model for the real number system.

https://doi.org/10.11647/OBP.0236.09

Given an infinite 'point set' P on the line (i.e. a subset of $\mathbb{R}$), he defined its *derived set* P' as the set of all its 'limit points'.[1] The point x belongs to P' precisely when infinitely many points of P *'lie within any neighbourhood, however small'* of x. This procedure can be iterated indefinitely, which led Cantor to his key result on Fourier series representations by defining two mutually exclusive 'species' of point sets. If, after n repetitions, the nth derived set $P^{(n)}$ is finite, then clearly its derived set $P^{(n+1)}$ is empty – he would then call P a set of the nth *kind*. A set P is of the *'first species'* if it is a set of the nth kind for some $n \geq 1$. Subsets of $\mathbb{R}$ whose derived sets $P^{(n)}$ were all infinite were placed in the *second species*.

This gave him the uniqueness criterion he was looking for: he showed that a real function f is represented uniquely by its Fourier series whenever, within any interval of length 2π, the set of exceptional points (where the series fails to converge or the representation fails) is a set of the first species.

But his investigations had now led him into quite different territory: Although the set of exceptional points is typically infinite, one might expect it to be 'small' compared to the set of points within the interval where the Fourier representation of f is unique. Therefore, in what sense could one distinguish between different 'sizes' of infinite sets? How might one extend the notion of counting to such sets? These questions about the nature of the continuum lay at the root of Cantor's extensive exploration of infinite sets.

1.1. Countably infinite subsets of $\mathbb{R}$. To pin down 'how many' elements a given set has, the natural first question concerns counting:

How should we count the 'number of elements' of a set?

Intuitively, a set A is finite precisely when we can 'count off its members one by one' in a list with a beginning and an end. In other words, there should be some natural number n such that we can 'pair off' all the elements of A one by one with the numbers $1, 2, 3, ..., n$. We would then say, quite naturally, that A has n elements, and write the list of its members as a *finite sequence*: $a_1, a_2, a_3...., a_n$.

Such a pairing is a *one-one correspondence* between the sets $\{1, 2, 3, ..., n\}$ and A.

In this way, n serves to tell us 'how many' elements the set A has. Cantor expressed this as the *power* ('Mächtigkeit') of the set A. Today we say that A has *cardinality* n, and write this as $|A| = n$. It is clear that no such finite pairing can exist for $\mathbb{N}$, which is therefore *not* a finite set.

Bolzano had made similar observations some 40 years before Cantor, and gave examples of infinite sets which he could put into one-one correspondence with proper subsets of themselves. But his notes came to light

---

[1]Now known as *accumulation points*. This notion was not new: it is implicit in *Bolzano's* version of the Bolzano-Weierstrass theorem, discussed in **Chapter 6.**

only much later. As Dedekind, independently, had also realised, the concept of one-one correspondences between sets leads naturally to a *definition* of infinite sets:

An *infinite set* is one that can be put into a one-one correspondence with a proper subset of itself.

Obvious examples are the set $\mathbb{N}$ of all natural numbers and the set $\mathbb{E}$ of all even numbers (where $n \in \mathbb{N}$ corresponds uniquely to $2n \in \mathbb{E}$); or the set of all perfect squares (where n corresponds to n^2); or the set of all prime numbers (although, in this example we can't locate the nth prime number precisely when n is large – as we saw in **Chapter 7.**)

Cantor observed that with one-one correspondences he could extend the notion of 'counting' to any set B that can be written as an infinite sequence $b_1, b_2, b_3, ..., b_n, ...$. Listing its elements as a sequence establishes a one-one correspondence (or *bijection*) between the set B and the set $\mathbb{N}$ of all natural numbers: for each $n = 1, 2, 3, ...$, associating the element b_n of B with the natural number n. Thus: we will call a set B *countably infinite* if it can be put in one-one correspondence with all of $\mathbb{N}$.

The term *denumerable* is often used instead, while a set is usually called *countable* if it is either finite or countably infinite. Cantor introduced the term 'countable' in 1883. Dedekind had called such sets *simply infinite*.

Cantor's initial interest was to examine familiar subsets of the continuum to decide whether they are countably infinite. A striking example is the use of his *first diagonal method* to show that $\mathbb{Q}$ is countably infinite.

Restricting to positive rationals ($\mathbb{Q}^+$), we illustrate how $\mathbb{Q}^+$ can be written as a sequence in the following diagram, where we imagine an infinite square array containing all positive fractions:

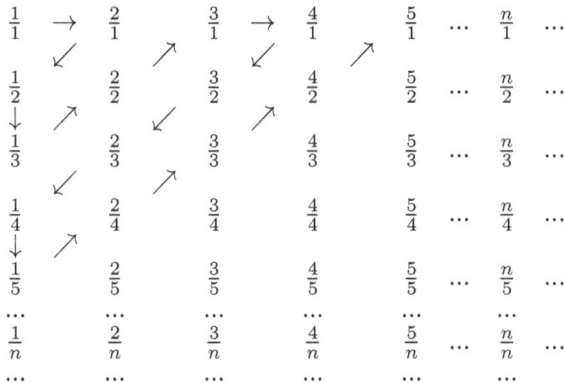

$$
\begin{array}{cccccccc}
\frac{1}{1} & \rightarrow & \frac{2}{1} & \frac{3}{1} & \rightarrow & \frac{4}{1} & \frac{5}{1} & \cdots & \frac{n}{1} & \cdots \\
\frac{1}{2} & & \frac{2}{2} & \frac{3}{2} & & \frac{4}{2} & \frac{5}{2} & \cdots & \frac{n}{2} & \cdots \\
\frac{1}{3} & & \frac{2}{3} & \frac{3}{3} & & \frac{4}{3} & \frac{5}{3} & \cdots & \frac{n}{3} & \cdots \\
\frac{1}{4} & & \frac{2}{4} & \frac{3}{4} & & \frac{4}{4} & \frac{5}{4} & \cdots & \frac{n}{4} & \cdots \\
\frac{1}{5} & & \frac{2}{5} & \frac{3}{5} & & \frac{4}{5} & \frac{5}{5} & \cdots & \frac{n}{5} & \cdots \\
\cdots & & \cdots & \cdots & & \cdots & \cdots & & \cdots & \\
\frac{1}{n} & & \frac{2}{n} & \frac{3}{n} & & \frac{4}{n} & \frac{5}{n} & \cdots & \frac{n}{n} & \cdots \\
\cdots & & \cdots & \cdots & & \cdots & \cdots & & \cdots &
\end{array}
$$

The arrows show how the *distinct* rational numbers in this array can be written as a sequence. From $1 = \frac{1}{1}$ move right to $2 = \frac{2}{1}$, then diagonally down left to $\frac{1}{2}$, down to $\frac{1}{3}$, then diagonally up right, (skipping $\frac{2}{2} = 1$)

to $3 = \frac{3}{1}$, then to $4 = \frac{4}{1}$, diagonally down left to $\frac{3}{2}, \frac{2}{3}, \frac{1}{4}$, down to $\frac{1}{5}$, then diagonally up right (skipping $\frac{2}{4}, \frac{3}{3}$ and $\frac{4}{2}$) to $5 = \frac{5}{1}$, then right to $6 = \frac{6}{1}$, diagonally left again, and so on, zig-zagging through the whole array. Every ratio whose numerator and denominator have common factors is skipped (recall that a rational number is an equivalence class of fractions). For example, $\frac{1}{5}$ becomes the tenth term in our sequence (as we skipped $\frac{2}{4}$), while $\frac{5}{1}$ is the eleventh, since $\frac{2}{4}, \frac{3}{3}, \frac{4}{2}$ are skipped. This ensures that only *distinct* positive rational numbers are counted. We map this sequence of distinct members of $\mathbb{Q}^+$ one-to-one to the even numbers $2, 4, 6, ..., 2n, ...$. Similarly, the set $\mathbb{Q}^- = \{-r : r \in \mathbb{Q}^+\}$ of negative rationals is mapped one–to–one to the odd numbers $3, 5, ..., 2n + 1, ..$, and 0 is mapped to 1. Taken together, these provide a one-one correspondence between $\mathbb{Q}$ and $\mathbb{N}$.

A slight extension of the above argument shows that *any* countable union of countably infinite sets is countably infinite. Write this union as a sequence of sets, then write each set in the union as a sequence, list them below each other as in our array, and move through the array in the zig-zig manner indicated, skipping all elements encountered at an earlier stage. This process rearranges the elements and displays the countable union of sequences as a single sequence, in one-one correspondence with $\mathbb{N}$. Thus: an infinite sequence of infinite sequences can be re-arranged into a single infinite sequence.

Cantor continued his explorations of subsets of $\mathbb{R}$ by showing that the set $\mathbb{A}$ of all real algebraic numbers (see **Chapter 8**) also countably infinite.

Recall that we may take $\mathbb{A}$ as the set of all real roots of polynomials with integer coefficients, that is, solutions of equations of the form

$$c_m x^m + c_{m-1} x^{m-1} + ... + c_1 x + c_0 = 0,$$

where $c_i \in \mathbb{Z}$ $(i \le m)$ and $m \in \mathbb{N}$. Recalling that $|c_i|$ denotes the *modulus* of the integer c_i, define the *height* of x as $h = m + |c_m| + |c_{m-1}| + ... + |c_0|$.

For any given h there are only finitely many polynomials with height h, since clearly this would require m and each $|c_k|$ to be at most h. Any polynomial of degree m with integer coefficients has at most m distinct real roots, so it contributes at most m distinct algebraic numbers, since its factorisation includes the product of $k \le m$ linear factors.[2] The factors $(x - \alpha_1)(x - \alpha_2)...(x - \alpha_k)$ yield real roots $\alpha_1, , , \alpha_k$. So any polynomial of height h will contribute only finitely many algebraic numbers. Since all the real algebraic numbers are found as roots of such a polynomial for some height h, we can now write them down as a sequence, beginning with height

---

[2]As noted in **Chapter 4**, in some cases we obtain quadratic factors that cannot be factorised further if we allow only real roots, such as $x^2 + 1 = 0$. These quadratic factors produce conjugate pairs of complex roots instead. In such cases $k = m - 2n$ for some $n \ge 1$.

2 (there are none of height 1).[3] Since each height contributes only finitely many new numbers, and all real algebraic numbers appear in the sequence, the set $\mathbb{A}$ of all real algebraic numbers is countably infinite.

1.2. Uncountable subsets of $\mathbb{R}$. Despite his success in defining the real number system, a natural question worried Cantor: since the rationals are *dense* in the reals, so that between any two real numbers at least one rational number (indeed, infinitely many) can be found, how might one characterise the apparent difference in 'size' (or plurality) between these two sets? Since $\mathbb{Q}$ has gaps while $\mathbb{R}$ does not, intuitively there appear to be many more reals than rationals. On the other hand, the sets $\mathbb{Q}$ and $\mathbb{N}$ are in one-one correspondence, yet there are also many more rationals than natural numbers.

In November 1873 Cantor wrote to Dedekind, asking whether, in his view, one-one correspondences could be found between $\mathbb{R}$ and $\mathbb{N}$. Dedekind replied that he could offer no evidence that such a correspondence would be impossible. But Cantor soon solved the problem in dramatic fashion in a paper that appeared in *Crelle's Journal* in 1874—this paper in effect launched set theory as a new subject. In this paper he proved that there can be *no* one-one correspondence between $\mathbb{N}$ and $\mathbb{R}$. (An infinite set would later be called *uncountable* if it could not be put into one-one correspondence with $\mathbb{N}$.)

Cantor's proof (which we summarise below) was criticised as less than convincing by some of his peers. Undaunted, he boldly emphasised the significance of his result as follows: *'Thus I have found the clear difference between a so-called continuum and a set of the nature of the entirety of the algebraic numbers.'*

Rather than reproduce Cantor's original proof, we consider a slight reformulation of his argument.[4]

Given an arbitrary sequence $(x_n)_{n\geq 1}$ of real numbers, construct nested closed intervals $([a_n, b_n]_{n\geq 1})$—that is, for each i, $[a_{i+1}, b_{i+1}]$ is a closed subinterval of $[a_i, b_i]$—each chosen to ensure that, for each $n \geq 1$, x_n is not in $[a_n, b_n]$. Thus $x_1 \notin [a_1, b_1]$, $x_2 \notin [a_2, b_2]$,..., $x_n \notin [a_n, b_n]$, ... For each n, the interval $[a_n, b_n]$ has been constructed to avoid the first n points of our given sequence. By one version of the completeness property of $\mathbb{R}$ (see Footnote 9 in **Chapter 8**) the sequence of nested closed intervals has non-empty intersection $I = \cap_{n\geq 1}[a_n, b_n]$. No x_n in our original sequence can belong to I, hence any member of I is a real number not in our sequence, so that $\mathbb{R}$ cannot be countable.

A different proof, outlined by Cantor in 1891, is more common in textbooks today. It rests on his *second diagonal argument*: for simplicity, restrict attention to the infinite decimal expansions whose integral part is 0. The

---

[3]For $h = 2$ we have $x = 0$ or $2 = 0$; the latter equation is false, so 0 is the only algebraic number for height 2. You may check that height 3 yields just 1 and -1; height 4 yields $-2, -\frac{1}{2}, \frac{1}{2}, 2$; and that height 5 provides $\sqrt{2}$ as one of the first irrationals in this sequence.
[4]See *Tim Gowers'* blog (https://www.dpmms.math.cam.ac.uk/~wtg10).

proof is by contradiction: if these expansions could all be written down in a
sequence, they would produce a doubly infinite array of the form

$$\alpha_1 = 0.a_{11}a_{12}a_{13}a_{14}a_{15}...$$
$$\alpha_2 = 0.a_{21}a_{22}a_{23}a_{24}a_{25}...$$
$$\alpha_3 = 0.a_{31}a_{32}a_{33}a_{34}a_{35}...$$
$$\alpha_4 = 0.a_{41}a_{42}a_{43}a_{44}a_{45}...$$
$$\alpha_5 = 0.a_{51}a_{52}a_{53}a_{54}a_{55}...$$

....

where all the a_{ij} are chosen from $\{0, 1, 2, ..., 9\}$, and the sequence $(\alpha_n)_{n \geq 1}$
would contain *all* infinite decimal expansions whose integral part is 0. To
avoid duplication in the list we will write all terminating decimals in their
'recurring nines' form. Now construct another infinite decimal expansion
$\beta = 0.b_1b_2b_3...$, where, for each $i \geq 1$, the digit b_i is chosen to be different
from a_{ii}, and the digits 0 and 9 are not used. For each i this leaves seven
alternative choices. This new expansion is different from each of the expansions in the sequence $(\alpha_n)_{n \geq 1}$, since it differs from a_1 in the first digit, from
a_2 in the second, and so on. None of the b_i are 0, so β cannot be a terminating
expansion that coincides with a 'recurring nines' entry α_n in the above list.
But that means that the list does not contain *all* infinite decimal expansions,
so our assumption that the set is countably infinite has led to a contradiction.

Hence the interval $[0, 1]$, and thus also the set $\mathbb{R}$ of all (non-terminating)
infinite decimal expansions, cannot be placed in one-one correspondence
with $\mathbb{N}$.

On the other hand, $\mathbb{R} = \mathbb{Q} \cup (\mathbb{R} \backslash \mathbb{Q})$ consists of all rationals together with
all irrationals. The union of two countably infinite sets is countably infinite,
and $\mathbb{Q}$ is countably infinite. Therefore, *the set of all irrationals is uncountable.*

As described above, Cantor's paper also verified that the set $\mathbb{A}$ of all real
algebraic numbers is countably infinite. Together with all transcendental
numbers it again makes up the (unccountable) set $\mathbb{R}$, so the set of transcendental numbers must be uncountable. In fact, as Cantor observes, his argument proves Liouville's claim that every interval contains infinitely many
transcendental numbers.

The fact that there are 'many more' real numbers that behave like π, e
or Liouville's constant L, may be somewhat disconcerting. We have to accept that we have no way of 'knowing' (individually) most of the numbers
that we represent by infinite decimal expansions. In fact, we cannot even *define* most of these decimal expansions in a meaningful way in finitely many
words. Using a finite alphabet, there are only countably many possible sentences that we can form to articulate the definition of any particular number
in words (or symbols) .

These facts about the 'familiar' continuum help to explain why Cantor's paper met with a hostile reception in some quarters, most notably from his former mentor *Kronecker*, who was to become a bitter enemy. Kronecker seems to have used his pre-eminent position in Berlin to block Cantor's ambition for a post in Berlin or Göttingen. Cantor also believed that Kronecker successfully dissuaded journal editors from accepting his papers for publication – this does indeed appear to be have been the case, for example, with the prestigious *Crelle's Journal*, where Kronecker was an editor.

2. Cantor's transfinite numbers

By 1880 Cantor had become deeply involved in his formulation of the general theory of sets, and in particular in his extension of the number concept to include *transfinite* numbers, i.e. numbers 'beyond the finite'. With this, Kronecker's hostility intensified: his motivation was his adamant refusal to accept any notions of the actual infinite having a place in mathematics. In his 'arithmeticisation programme' Kronecker had insisted that *all* mathematics should be capable of being based on a finite number of operations with integers. In his 1886 article 'Über den Zahlbegriff' (On the number concept) he objected strenuously to many widely accepted mathematical developments of his time, such as the Bolzano-Weierstrass theorem, claims for the existence of suprema and infima, and even the irrational numbers. For example, having read Lindemann's proof that π is transcendental, he commented:

Of what use is your beautiful investigation of π. Why study such problems when irrational numbers do not exist.

From this extreme perspective it is no surprise that he saw Cantor's work as anathema. Sadly, the ensuing controversies and professional disappointments this created probably contributed to Cantor suffering a series of mental breakdowns that greatly hampered his mathematical research for significant periods. Ironically, the origin of the controversy is found in the notion of *counting*, with which I began this book, and which forms the basis of Pythagorean mathematics as well as of Kronecker's own position.

Cantor first perceived a need to extend the process of counting 'beyond the finite' while dealing with the structural analysis of linear point sets through his concept of derived set. His conclusions were published as a series of six papers entitled *"Uber unendliche lineare Punktmannigfaltigkeiten* (On infinite linear point sets) that appeared between 1879 and 1884.[5] One starting point was his classification of point sets of the first and second species,

---

[5]For Cantor's original papers see his annotated collected works, edited in 1932 by *Ernst Zermelo* [46]. Zermelo comments that the 'germ' of Cantor's theory of transfinite cardinals can be found in the transfinite sequence of derived sets.

where, as shown above, the first species could be divided into sets of the nth *kind* for $n \geq 1$.

As yet, he had defined no such subdivisions for the second species. In the second paper of his series he observed that, in the sequence of derived sets $P', P'', P''', \ldots$ of a set P, each member is a subset of the previous one. When P is a set of the second species, *its* derived set, again denoted by P', can therefore be written as a disjoint union $P' = Q \cup R$, where Q consists of all points 'lost' when we construct a sequence of successive derived sets of P (i.e. given $x \in Q$, there is a smallest $n \geq 1$ such that x does not belong to $P^{(n)}$), while $R = \cap_{n \geq 1} P^{(n)}$ consists of all points that belong to every derived set $P^{(n)}$ for $n \in \mathbb{N}$. Since P is not a set of the first species, he knew that $R \neq \varnothing$ and he denoted R by $P^{(\infty)}$, as the 'derived set of order ∞'.

The derived set of $P^{(\infty)}$ would then be denoted by $P^{(\infty+1)}$. He defined the next derived set as $P^{(\infty+2)}$, and continued in this fashion, denoting the nth-order derived set of $P^{(\infty)}$ by $P^{(\infty+n)}$. In this way, $P^{(\infty)}$ will also have a 'derived set of order ∞' consisting of points belonging to every $P^{(\infty+n)}$. This set would be denoted by $P^{(2\infty)}$. Continuing in this fashion, Cantor constructed the derived set of order '$n_0\infty + n_1$' for any natural numbers n_0, n_1. This led to the next 'limit set' as the set $P^{(\infty^2)} = \cap_{n \geq 1} P^{(n\infty)}$, and then to 'polynomial combinations' of the symbol ∞ in the form $n_0\infty^\nu + n_1\infty^{\nu-1} + \ldots + n_\nu$. (Note that this was simply his *notation* to identify the 'positions' of sets in the list, without implying 'arithmetical' operations!) Treating the power ν as a *variable*, this leads, as before, to $P^{(\infty^\infty)} = \cap_{\nu \geq 1} P^{(\infty^\nu)}$. This, in turn, generated new derived sets and the process could be continued indefinitely.

Although at this stage the infinite symbols served primarily as *labels* by which he could distinguish between various *levels* of derived sets, Cantor stated boldly that his successive definitions amounted to a '*dialectical generation of concepts, which continues ever further and, free of any arbitrariness* [Willkür], *remains consistent and necessary in itself*'.

2.1. Cardinal numbers. Thus the roots of Cantor's investigations of different types of infinite sets are illustrated by two distinct aspects of his early work described above:

(i) his discovery that some familiar infinite sets are of different 'sizes';

(ii) his classification of sets of the first and second species (via their sequences of derived sets).

We now consider where these led him in turn.

Cantor's definition of the *cardinal number* (or *power*) of a set was to be of crucial importance, extending the idea of one-one correspondences from $\mathbb{N}$ to sets in general. Readers familiar with modern set theory will be aware

that the definition of this concept has become considerably more sophisticated since Cantor's time, depending on the particular axiom system employed to define the notion of *set*. We will avoid such issues here and restrict
our attention essentially to Cantor's perceptions. What remains basic to our
(naive) setting is the following:

Definition

Two sets M, N are *equipotent* (also called *equipollent*) if there is a one-one
correspondence (also called a bijection) between them. In that case M and
N are said to have the same *cardinal number* (or *power*) and we denote this
by $M \sim N$.

Cantor argued that this concept is not restricted to sets of whole numbers, but *'ought to be considered as the most general genuine foundation* [he used
the German word 'Moment'] *of sets'*. Here we see him already claiming *set
theory* as the basis of all 'pure' (or, as he would have it, *'free'*) mathematics.
In his later papers (1895/1897) he used the notation $\overline{\overline{M}}$ to denote the power
of the set M, arguing that it was *'that general concept which, with the help of
our active thought-processes, arises from the set M, abstracting from the character
of its various elements m and from the order in which they occur'*. Statements
such as these reveal a philosophical stance that asserts the reality of mental
constructs and the primacy of consciousness.

I will not use Cantor's notation, but simply write $|M|$ to denote the car-
dinal number of M. For the purposes of discussing its cardinality M can be
replaced by any set equipotent to M. The cardinality of a set has nothing to
do with any 'ordering' of its elements.

At this stage Cantor only had two distinct infinite examples (sets equipotent with either $\mathbb{N}$ or $\mathbb{R}$), but he asserted confidently that the concept of
power *'is by no means restricted to linear point sets, but can be regarded as an at-
tribute of any well-defined collection, whatever may be the character of its elements'*.
He was soon to provide a more detailed justification of this claim.

In 1883 he published the fifth article in his series on point sets in the
journal *Mathematische Annalen*. This was also published separately and is
now known as his *Grundlagen* [Basics] paper. Here he developed his ideas
about infinite sets in an abstract setting, using his previous results about derived sets as a guiding example. He prefaced the presentation of his theory
with an extensive discussion of philosophical and historical objections to
the concept of *actual infinity*, a concept which he saw as central to his whole
project.

His research in set theory had reached a point where further progress
would depend upon a systematic *'extension of the concept of real whole number
beyond its previous boundaries'*, and this extension had taken a direction that,
to his knowledge, no-one had taken before him. Although he expressed the
hope, indeed firm conviction, that the idea of the actual infinite would *'in*

time, have to be regarded as a thoroughly simple, appropriate and natural one', he was well aware that he was placing himself *'in a certain opposition to wide-spread views about the mathematical infinite and to frequently advanced opinions on the nature of number'*.

He pointed out that Aristotle's notion of the *potential infinite* had habitually been used to justify the Calculus. This relied crucially on the idea of variable finite quantities that could grow or shrink beyond any assignable bounds, while remaining finite at any particular stage. Cantor would now refer to such notions as describing *improper infinities*. By way of contrast, the widely accepted practice of postulating the existence of a *'point at infinity'*, which was prominent in both projective and hyperbolic geometry and via the Riemann sphere in complex function theory, had a quite different character.[6] These were fixed *ideal* points, justifying their description as actual, or as he wished to call them, *proper infinities*.

2.2. Ordinal numbers. The *transfinite numbers* he would now define had the latter character. He had worked with them for some years without fully realising that they constituted *'concrete numbers of real meaning'*. In contrast to a single ideal 'point at infinity', he would introduce, successively, an infinite collection of infinite numbers, all differing from one another. Their construction would be based on two distinct *principles of generation* together with a *limitation principle* that would serve to distinguish between different classes of numbers within this collection.

Thus Cantor was ready for the second fundamental innovation suggested by his earlier work on sets of limit points: he would extend the finite *ordinals* (or *ordinal numbers*), indicated by their position in the sequence $1, 2, 3, ..., n, ...$, indefinitely beyond the finite. These ideas constitute his second major breakthrough, initiating an entirely new subject of study.

In this area also, mathematics has evolved substantially since Cantor's time, but his basic ideas largely remain intact. We will see below how—as Cantor himself described in his later papers—the concept of *well-ordering* (which, for $\mathbb{N}$, follows from the induction principle—see **Chapter 7)** was to become fundamental to his theory.

Today, one common procedure is to identify an ordinal as the set of all ordinals that precede it in the given ordering – as is done for finite ordinals in von Neumann's model for $\mathbb{N}_0$ in **Chapter 7**. Thus ordinals 'label' the positions of elements of a set, whose *order type* is then given by the least ordinal that is not a member of that set. By way of contrast, *cardinals* only express

---

[6]For the last of these, imagine 'bending' the Gaussian plane into a sphere, so that the origin 0 forms the South Pole, while the boundaries of the four quadrants meet at the North Pole, which we then treat as the 'point at infinity', denoted by ∞. Points in the plane with very small coordinates map to points near 0 while points with very large coordinates map to points 'near' ∞.

the 'size' of the set. As Cantor pointed out in the construction we describe below, many different infinite ordinals will have the same cardinal number, since the latter takes no account of the ordering of the elements of two sets being compared, only the existence of a one-one correspondence between them.

Cantor's starting point was to consider how counting, i.e. starting at the unit 1 and successively adding a unit each time, enables one to create the set of all natural numbers, which he denoted by (I) (the 'first number class') for this purpose. Its elements were the *finite ordinal numbers*. Although it would be contradictory to speak of a 'largest number' in this set, nothing prevented him from *defining* a new number ω that expressed the natural, regular order of the set (I) as a whole.[7] The symbol ω would represent the first *transfinite ordinal* number, the first number to follow the entire sequence of natural numbers ν. He argued that it was legitimate to think of ω as a 'limit' to which the natural numbers ν 'tend', provided that we meant by this that ω should be the *first* number that follows all the natural numbers.

But, having defined ω, he could now continue adding units successively, creating new transfinite ordinal numbers

$$\omega + 1, \omega + 2, ..., \omega + \nu, ...$$

again producing a new sequence without a largest element. Nonetheless, the ordinals $\omega + \nu$ ($\nu \in \mathbb{N}$) are all equipotent to ω (as sets!) so they all have the same *cardinal* number. Applying the same logic as when discussing derived sets, he defined a new transfinite ordinal number 2ω as denoting the set consisting of all numbers of the form ν or $\omega + \nu$, with ν taken from (I).[8] As he had done for his derived sets, Cantor now repeated the use of his two generating principles—in describing this we use his original notation. The first principle is the *successive addition of units*, while the second only comes into play for a '*definite succession of defined whole numbers...for which there is no largest*'. In such situations the new number created is the '*next number larger than all of them*', the first two examples being ω and 2ω. Using these two principles repeatedly, he could first reach transfinite numbers of the form $2\omega + \nu$ for all ν in (I), to be followed immediately by 3ω, then all $3\omega + \nu, ..., \mu\omega + \nu$, etc., so that the ordinal immediately following all these is

---

[7]He commented in a footnote that he would now use ω rather than ∞, precisely because the latter symbol was frequently used to signify the *potential* infinite (as in $x = \lim_{n\to\infty} x_n$) rather than denoting an *actual* infinite number, as was required here.

[8]In the 1890s Cantor use the notation $\omega 2$ instead of 2ω, and this is the notation used today. One way of envisaging $\omega 2$ (or $\omega + \omega$) is as *two copies* of ω, (e.g.) representing the infinite sequence

$$1, 3, 5, 7, 9, ...; 2, 4, 6, 8, 10, ...,$$

since, just like the set used to define $\omega 2$, it has two numbers (1 and 2) which are not immediate successors of any number, and each of the sequences $1, 3, 5, ...$ and $2, 4, 6, ...$ can be put in one-one correspondence with $\mathbb{N}$. What matters here is the *order structure* rather than any specific 'labels' (including any 'arithmetical' notation) used to identify individual elements.

denoted by ω^2. Using the symbol $+$ to indicate how the successions proceed in each case, he could describe all transfinite numbers of 'polynomial' form as $\nu_0\omega^\mu + \nu_1\omega^{\mu-1} + ... + \nu_\mu$, where μ, ν_k are natural numbers, while the collection of all of these would be followed by ω^ω, and so on!

Today, ordinal numbers such as $\omega, \omega 2$ or ω^ω, which are neither 0 nor *successor ordinals* (i.e. produced by adding 1 to an earlier ordinal) are called *limit ordinals*.

His formulation led Cantor to a *limitation principle* ['Hemmungsprinzip'], whereby he would identify breaks in the seemingly endless process of number creation: repeating the essence of the proof that the set $\mathbb{A}$ of all real algebraic numbers is countable, the above collection of numbers of 'polynomial' form is shown to be countable, as μ, ν_k are natural numbers.

In his hierarchy of number classes, class (I) comprises the natural numbers. He now defined the second number class, denoted by (II), as:

'the collection of all numbers, increasing in definite succession, which can be formed by means of the two principles of generation:

$$\omega, \omega + 1, ..., \nu_0\omega^\mu + \nu_1\omega^{\mu-1} + ... + \nu_\mu, ..., \omega^\omega, ..., \alpha, ...$$

subject to the condition that all numbers preceding α (from 1 on) constitute a set of the power of the first number-class (I).'

In other words, 'initial segments' of the second number class (II) were to remain countable sets, just as initial segments of class (I) were finite. Cantor went on to prove that number class (II) has a higher power than number class (I), and also that this power immediately follows that of the first.[9] For the last of these claims he needs to make use of the smallest ordinal number in class (III). This is the only occasion (in the *Grundlagen*) where he mentions the third number class (III): it consists of all numbers 'generated' by repeating the above process, starting with number class (II) instead of with class (I). Nonetheless, he asserts without further ado that the process of generating new transfinite numbers can be continued indefinitely, creating an unlimited collection of number classes, each subject to his limitation principle, which in its general form states that new transfinite numbers can be created by use of the two generating principles *'only if the totality of all preceding numbers has the power, in its whole extent, of an already defined number class'*.

---

[9]Cantor's proofs of these claims are quite unwieldy, but already contain the seeds of the notion of *well-ordering*, which became the centrepiece of his later reformulation of the theory of transfinite numbers. He published his new approach only in 1897. In the *Grundlagen* Cantor simply states without proof that any non-empty sub-collection of the collection of all ordinals has a first element. In [46] his editor *Zermelo* provides a straightforward proof of this fact.

Basing his treatment on the *order* concepts he had introduced, he ended the paper by developing the *arithmetic* of transfinite ordinals in considerable detail, showing that their algebraic properties are quite different from those elaborated earlier for $\mathbb{N}$, $\mathbb{Z}$ and $\mathbb{Q}$. In particular, the *commutative laws* for addition and multiplication break down even in simple cases. For example, $2 + \omega \neq \omega + 2$. To see why this is so, note that we can write these two sets as follows:

$$2 + \omega = (1, 2, a_1, a_2, ..., a_\nu, ...),$$
$$\omega + 2 = \{a_1, a_2, ..., a_\nu, ...; 1, 2\}.$$

These two sets are equipotent, but are not equal as *ordinal numbers*, as the orderings do not correspond: in the first, only one element, 1, has no immediate predecessor; in the second there are two such elements, a_1 and 1. Similar arguments show that multiplication is not commutative in general: for transfinite ordinals α, β we find that $\beta\alpha \neq \alpha\beta$, since α copies of β need not have the same order structure as β copies of α. But we will not delve further into the arithmetic of transfinite ordinals here.

3. Comparison of cardinals

A widely used notation for the *cardinality* of the various number classes, introduced by Cantor in the 1890s, uses the Hebrew letter $\aleph$ (*aleph*). It lists the cardinality of the first number class as $\aleph_0$ and that of the second number class as $\aleph_1$. The precise relationship between $\aleph_1$ and the cardinal number of the *continuum* was to occupy much of his subsequent work. Recall that in the *Grundlagen* he had shown that the cardinalities of his number classes (I) and (II) were distinct. He had also claimed correctly that the cardinality of number class (II) was the next greatest after that of class (I) in this sequence. Here his arguments, based on what *Ernst Zermelo* (1871-1953) describes as a *'purely constructive'* definition of the two 'generating principles', lacked much of the clarity of the treatment provided when Cantor revisited the matter in his *Beiträge* papers in 1895 and 1897. The claim that the alephs constitute an infinite number of *distinct* cardinal numbers would need further clarification.

A fundamental question concerned the *comparability* of transfinite cardinal numbers. For any two *distinct* finite numbers m, n we know that either $m < n$ or $n < m$ will hold; this is what we called the *trichotomy* for the (total) ordering of $\mathbb{N}$. To extend this to transfinite cardinals (and thus to justify their designation as 'numbers'), Cantor suggested in 1887 that, given two sets M, N, the inequality $|M| < |N|$ should mean that there is a proper subset N' of N that is equipotent to M, while no subset of M is equipotent to N. This ordering of cardinal numbers is easily shown to be transitive (see *MM*).

Cantor proved that *at most one* of $|M| < |N|, |M| = |N|, |N| < |M|$ can hold. This is straightforward: by definition equality cannot hold at the same time as either of the other relations. But if $|M| < |N|$ then some $N_1 \subset N$ is equipotent to M, which means that we cannot have $|N| < |M|$.

However, the same could not be said for the claim that *at least one* of the above relations must hold for arbitrary sets M, N. For given M, N there are two further possible outcomes in addition to the relations $|M| < |N|$ and $|N| < |M|$:

(i) M is equipotent to a subset of N and N is equipotent to a subset of M,

(ii) M is equipotent to no subset of N and N is equipotent to no subset of M.

Cantor claimed that case (i) would ensure that M and N are equivalent, but he never proved this. It was proved, independently in 1897, by *Ernst Schröder* (1841-1902) and by his student *Felix Bernstein* (1878-1956), who had corrected an error in Schröder's original claim of this result, published in 1896.

The Schröder-Bernstein theorem:

If each of M, N is equipotent to a subset of the other, then M is equipotent to N.

Although the proof of this theorem does not require advanced tools, it is by no means obvious (see *MM*). It eluded Cantor until after his principal papers on set theory had been published.

Case (ii) above implies that M and N are not comparable by the relation $<$, which would mean that it is not a total order. Initially, Cantor was unable to exclude this possibility for the cardinals of infinite sets. However, in his *Beiträge* papers of 1895/97 he provided a complete reformulation of his number classes, based on the concept of a well-ordered set, instead of the more nebulous 'generating principles' presented in the *Grundlagen*.

The modern definition echoes and generalises the Well-Ordering property (WO) proved in **Chapter 7** for $\mathbb{N}$:

A set M is *well-ordered* in a given ordering if every non-empty subset of M has a first element in that ordering.

Cantor's definition was more elaborate. He first defined a set as *simply ordered* if for any two of its members one can always be shown to precede the other. Two simply ordered sets M, N are *similar* if there is a one-one correspondence ϕ between them that respects order, i.e. (denoting their orderings by $<_M, <_N$) if $m_1 <_M m_2$ then $\phi(m_1) <_N \phi(m_2)$. The two sets are said to have the same *order type* – Cantor wrote this as $\overline{M} = \overline{N}$ – if and only if they are similar[a].

He distinguished between *number* [Zahl] and *numbering* [Anzahl]. The former relates only to the size (i.e. cardinality) of the set, the latter takes the ordering of the elements into account, insisting that the one-one correspondence between the sets should preserve the ordering. For finite sets, of course, the two notions coincide, so this distinction would suffice to characterise actual infinite sets. He argued that the centuries-old confusion about potential and actual infinities might have had its origin in the fact that finite numbers function in this dual sense.

The upshot of his reasoning was that the first transfinite number $\aleph_0$ could be taken as that of the first number class (I), in other words, the ordinal ω. The second number class (II) was defined as *'the entirety of all order types α of well-ordered sets of cardinality $\aleph_0$'.* By showing that this is a well-ordered set, he could define the second transfinite number $\aleph_1$ as its least element, and prove the inequality $\aleph_0 < \aleph_1$.

---

[a]In much the same way as for cardinal numbers, we would today define an order type μ as any representative of a class of mutually similar sets. Clearly two sets with the same order type define the same cardinal number, but the converse is false in general.

It is clear that well-ordering is an intrinsic part of any counting procedure, as we saw when discussing $\mathbb{N}$. By 1883, Cantor had become aware of the centrality of well-ordered sets for his entire set theory, but he did not prove that his set of transfinite cardinals could be well-ordered. Instead, in the third section of the *Grundlagen* he made the claim that *'any well-defined set can be brought into the form of a well-ordered set'.* He regarded this as *'a basic law of thought with far-reaching consequences especially remarkable for its general validity'* to which he promised to return in a later paper. However, by the 1890s he had realised that his bold claim was by no means self-evident. This led him to a thorough reformulation of his transfinite ordinals, published in Part II of the *Beiträge* (1897), and at last enabled him to resolve the awkward question of the comparability of his alephs.

3.1. Cantor's second diagonal argument. The publication of Cantor's *Beiträge* in 1895 and 1897 met a more receptive audience than had his earlier work in the *Grundlagen*. His primary critic, Kronecker, had died in 1891, and the younger generation of mathematicians throughout Europe showed greater willingness to grapple with the fundamental questions Cantor's work had raised. The *Beiträge* were soon translated widely. They also proved to be more accessible, providing firmer foundations for some of Cantor's claims that had been the case in earlier work.

Opposition to Cantor's ideas had not gone away, however. For example, the great French mathematician, *Henri Poincaré* (1854-1912) remained a stern and influential critic of transfinite numbers, calling the theory a 'disease' from which mathematics would eventually recover!

In the late 1880s, disappointed at his failure to obtain the recognition and prestigious position he had hoped for, Cantor was active in campaigning for a new professional body for German mathematicians. The established professional organisation, representing mathematics and medicine,

seemed to him moribund, personifying the academic establishment that
had blocked his publications and career aspirations. His advocacy of an alternative resulted in the formation of the *Deutsche Mathematiker-Vereinigung
(DMV)* [German Mathematicians' Union] which elected Cantor as its first
President at its inaugural meeting, held in Halle in 1891.[10]

Cantor used this occasion to present what has become one of his most
distinctive and important contributions. We have already seen an application of his *'second diagonal argument'* in the proof of the uncountability of
the reals. His own description of his simple, yet groundbreaking technique,
published under the unassuming title: *'Über eine elementare Frage der Man-
nigfaltigkeitslehre* (On an elementary question in set theory) and taking up
just three pages of the first volume of the DMV's annual reports, makes interesting reading. As he pointed out, this proof was the first to be entirely
independent of the definition of irrational numbers, and lent itself to a vast
range of generalisations.

He began with just two distinct elements, m and w, and considered the
set M of all possible sequences $E = (x_i)_{i\geq 1}$ such that each x_i is either m or w.
(In these binary days of computer science, we would immediately translate
these into sequences using only 0 and 1.) If the set M of these sequences
were countable, we could write it as a sequence, so its elements could be
listed as

$$E_1 = (a_{11}, a_{12}, ..., a_{1n}, ...)$$
$$E_2 = (a_{21}, a_{22}, ..., a_{2n}, ...)$$
$$....$$
$$E_n = (a_{n1}, a_{n2}, ..., a_{nn}, ...)$$
$$....$$

The sequence $(b_1, b_2, ..., b_n, ...)$, where, for each $n \geq 1$, b_n is either m or
w, but where we insist that $b_n \neq a_{nn}$, is obviously a member of M but does
not equal any of the E_i. Thus M cannot be countable.[11]

Cantor showed that the diagonal argument can be applied to any set M
to show that the cardinality of a set is *always* less than that of its so-called
power set, $\mathcal{P}(M)$, defined as the set of *all* subsets of M (including $\varnothing$ and
M itself). To see why, let us begin by counting the subsets of small sets:

---

[10]The DMV remains the premier professional organisation for German mathematicians
today.

[11]Cantor added that the same technique can be used to prove the uncountability of $\mathbb{R}$.
As *Zermelo* remarked in a footnote when editing Cantor's Collected Works in 1932, this claim
needs a minor amendment: binary expansions do not represent rational numbers uniquely,
since expansions of the form $0.a_1a_2...a_n01111...$ and $0.a_1a_2...a_n10000...$ represent the same
rational in $[0, 1]$, for example. But we can always decide in advance which representation to
use throughout – exactly as we did for decimal expansions in **Chapter 7**, Section 6.3.

Figure 39. Henri Poincaré sitting (Henri Manuel)[12]

$\varnothing$ has only one subset, namely itself,

the singleton set $\{m\}$ has two subsets, $\varnothing$ and $\{m\}$,

the set $\{a, b\}$ has four, namely $\varnothing, \{a\}, \{b\}, \{a, b\}$,

the set $\{a, b, c\}$ has eight: $\varnothing, \{a\}, \{b\}, \{c\}, \{a, b\}, \{a, c\}, \{b, c\}, \{a, b, c\}$.

In general, a set with n elements has 2^n subsets. This follows from the binomial theorem: $(a + b)^n = \sum_{k=0}^{n} \binom{n}{k} a^k b^{n-k}$ (see **Chapter 5**), taking $a = b = 1$; a set with n elements has $\binom{n}{k} = \frac{n!}{k!(n-k)!}$ distinct subsets with k elements.

So what about infinite cardinals? Write the power set of M as $\mathcal{P}(M)$. The above suggests the *notation* $|\mathcal{P}(M)| = 2^{|M|}$. Cantor's argument showed that $|M| < |\mathcal{P}(M)|$, and this immediately provides an infinite, strictly increasing sequence of infinite cardinal numbers. (The proof is given in *MM*.)

In terms of Cantor's aleph notation for infinite cardinals, this means that the power set $\mathcal{P}(\mathbb{N})$ has a higher cardinal number greater than $\aleph_0$. By analogy with a set with n elements, whose power set has 2^n elements, we may adopt the *notation* $2^{\aleph_0}$ for the cardinality of the power set $\mathcal{P}(\mathbb{N})$.

3.2. Unsolved problems and paradoxes. Despite wrestling with it for many years, Cantor remained unable to resolve a fundamental question that

---

[12]https://commons.wikimedia.org/wiki/File:Henri_Poincaré_sitting.jpg

had occupied him since the late 1870s: given that the real number system $\mathbb{R}$ is uncountable, is its cardinal number the *next greatest* after that of $\mathbb{N}$? Although he could not prove this, Cantor remained convinced that it is, and this claim became known as his *Continuum Hypothesis (CH)*.

With the notation developed above, Cantor's Continuum Hypothesis (CH) can now be framed succinctly. He knew that the real numbers can be placed in a one-one correspondence with the power set $\mathcal{P}(\mathbb{N})$ of the natural numbers.[13] Denoting the cardinality of the real number system $\mathbb{R}$ by c, this means that $c = 2^{\aleph_0}$.

Cantor's claim is that there is *no* set with cardinal number *strictly* between $\aleph_0$ and c. In other words, his Continuum Hypothesis asserts that any subset X of $\mathbb{R}$ is *either* countable *or* has $|X| = c$. In the *well-ordered* sequence of transfinite cardinals $\aleph_1$ is the next greatest cardinal after $\aleph_0$. Thus the Continuum Hypothesis takes the form: $\aleph_1 = 2^{\aleph_0}$.

In this form, Cantor's hypothesis can be generalised in terms of an *arbitrary* infinite cardinal λ. The *Generalised Continuum Hypothesis* (GCH) states that there can be no infinite cardinal lying between λ and 2^λ. In terms of ordinals and alephs it then reads: for any ordinal α, $\aleph_{\alpha+1} = 2^{\aleph_\alpha}$.

In two critical aspects, therefore, Cantor's hopes to provide a secure basis for all of set theory were not realised:

(a) He had made no real progress on the question whether every set can be well-ordered, although this claim remained fundamental to his theory.

(b) He had not been able to prove his Continuum Hypothesis.

Moreover, he had become aware that the concept of the set *all* cardinals, or that of all ordinals, appeared *self-contradictory* if these were also to be considered as 'sets'—in other words, his set theory contained *paradoxes*.

The earliest paradoxes arose when basic questions were asked about the nature of the collection of '*all*' objects of a particular kind.

(i) The simplest paradox, named after Cantor, questions whether *the set of all sets*, S, can be a set. If so, it must equal its power set $\mathcal{P}(S)$: if S is a set, then $\mathcal{P}(S)$ is also a set, and by definition it is both contained in (the set of *all* sets) S and contains $\{S\}$ as an element. This, however, yields the contradiction $|\mathcal{P}(S)| = |S| < |\mathcal{P}(S)|$ by the above diagonal argument, and therefore shows that S *cannot* be a set. In other words, the process of set formation

---

[13]Essentially, map $S \subset \mathbb{N}$ to an infinite binary sequence $0.a_1a_2...a_n...$, using $a_n = 1$ if $n \in S$ and 0 otherwise. This maps subsets of $\mathbb{N}$ injectively into binary representations of real numbers in $[0, 1]$. This will imply that $c \geq 2^{\aleph_0}$. On the other hand, treating real numbers as Dedekind cuts (i.e. subsets of $\mathbb{Q}$) means that c is no greater than $|\mathcal{P}(\mathbb{Q})| = |\mathcal{P}(\mathbb{N})| = 2^{\aleph_0}$. So by the Schroeder-Bernstein theorem, $c = 2^{\aleph_0}$.

without *any* limitations appears to be highly problematic. Cantor was probably aware of this paradox in 1895, and certainly before he published Part II of his *Beiträge* in 1897.

(ii) Cantor was aware of a similar result announced in 1897 by Peano's former student *Cesare Burali-Forti* (1861-1931). This was the first paradox of set theory to be published. It arises when we consider the set Ω of *all ordinals* (recall that ordinals are themselves sets). Now, if Ω is a set, then we can, according to Cantor's prescription, form its successor ordinal, which we would denote by $\Omega + 1$. But, as before, we would obtain the nonsensical inequalities $\Omega < \Omega + 1 \leq \Omega$. So, the 'set of all ordinals' is also a meaningless concept. Burali-Forti's paper did not arouse much interest at first, nor was much concern expressed when similar arguments showed that the set of all cardinal numbers, or indeed, the set of all alephs, led to similar contradictions.

To deal with these questions, Cantor sought to distinguish between what he called 'consistent' and 'inconsistent' concepts. He wished to treat the former as sets, but exclude the latter as 'absolutely infinite', which, he argued, *'can never be conceived complete and actually existing'*. To describe this distinction he began to formulate *axioms* that the process of set-formation would need to satisfy.

There was by now a wider recognition that it was Cantor's very general definition of what constitutes a *set* (as given in the *Beiträge*) that would lead to *paradoxes* (logicians prefer to call them *antinomies*, i.e. real contradictions that can be deduced by applying specified logical rules to an apparently true claim). Dedekind's notion of *infinite systems*, which he espoused in *Was sind und was sollen die Zahlen?* as an alternative way of describing sets in general, would lead to similar conclusions.

The task of avoiding antinomies was later taken up by Bertrand Russell who argued that, instead of Cantor's 'inconsistent' entities, one should consider *properties which do not determine a set* (that is, there is no set consisting exactly of the objects that have the property). This conceptual shift, towards describing mathematical entities by means of logical concepts, as well as the search for an axiomatic basis of set theory, was to become a key element of research for several decades, and led to much of the modern subject of mathematical logic.

One particular intervention by Russell was soon to complicate matters further. The catalyst was a letter (dated 16 June 1902) from Bertrand Russell to the German logician and mathematician *Gottlob Frege,* who had just completed the second volume of his major work *Grundgesetze der Arithmetik* [Basic Laws of Arithmetic]. Russell's letter led Frege to the conviction that the edifice he had built over a lifetime contained a fundamental flaw. Later he said that the paradox that Russell had discovered had destroyed set theory! To understand why, we need to outline the background and nature of Frege's own investigations.

The purpose of Frege's research had been to base arithmetic upon purely logical concepts. This programme to derive all mathematical principles from the laws of logic alone, became known as *logicism*. It had attracted mathematical philosophers, including Russell, as well as other mathematicians such as Dedekind and Peano.

In philosophical terms the logicist programme opposed the materialism of *David Hume* and *John Stuart Mill*, who argued that our mathematical ideas ultimately arise from our senses through observation. At the same time the logicist viewpoint opposed *Kant's* notions of our *a priori* intuitions of space and time. For example, in the Preface to his *Was sind und was sollen die Zahlen?*, published in 1888, Dedekind had located his concept of number firmly within *'the laws of thought'*; unlike Hamilton, who had earlier attempted to describe number (and algebra) as reflecting our *a priori* intuition of 'pure time'.

In his *Foundations of Arithmetic* [Grundlagen der Arithmetik], published in 1884, *Frege* had addressed many of the same issues as Dedekind, but as seen from the viewpoint of a logician rather than as a mathematician. He had developed a meticulous language to express logical concepts, rules of inference and logical axioms. This served to clarify the nature of mathematical reasoning and set the stage for what is known as *predicate calculus* in mathematical logic today. In his setting, a *mathematical proof* is a finite sequence of statements, each of which is either an axiom or follows from previous statements in the sequence verified by valid rules of inference.

Frege's notation and mode of argument would take us beyond the scope of this book, but we can indicate why Russell's letter had such a destructive impact upon Frege's system. His two-volume *Grundgesetze der Arithmetik* (1893/1903) sought, as the title suggests, to identify a small number of 'basic laws' of arithmetic upon which the whole structure could be erected solely through the use of logical terms and rules of inference. A key logical axiom Frege needed to complete his programme was his *Basic Law V*.[14] Rather than use Frege's abstruse terminology and notation, we explain the difficulty in terms of the (implicit) assumptions about set-formation used by both Cantor and Dedekind, which allowed the formation of sets through *self-referential* concepts.

---

[14]In his logical universe, containing only objects and functions (the latter taking an object to a value), Frege wished to express the notion of the extension (he used the German word Umfang) of an (unspecified) object. The extension of a concept F records the objects for which F holds. However, Russell realised that, under Basic Law V one can form a self-contradictory concept, by defining x as the extension of some concept which does not apply to x.

Cantor had argued that the term *set* should apply to '*every gathering together into a whole of definite, distinct objects m of our perception or of our thought*', while Dedekind said that '*different things...can be considered from some common point of view, can be associated in the mind, and we say that they form a system S*'. Frege criticised these statements, which essentially contend that 'any precisely specified property' will suffice to define a set by stipulating the conditions for membership of the set.[15] But despite Frege's careful construction of his logical system, his Basic Law V in effect amounts to making a similar claim, as Russell pointed out. We can see how Russell created a self-contradictory set under these assumptions:

If we denote *the collection of all sets that are not members of themselves* by R, then, according to Frege's Basic Law V (reformulated in terms of sets), R is admissible as a set. But now we cannot answer the question whether R is a member of itself! For, if we have $R \in R$ then, by definition of R, we must have $R \notin R$. On the other hand, if $R \notin R$ then, again by definition of R, it follows that $R \in R$. This then was Russell's Paradox, which sent the whole logicist programme into considerable turmoil.[16]

It seemed that, in order to maintain the freedom obtained by working with sets in general, contradictions could only be avoided if careful limits were placed on the process of set formation. Set theory urgently required a *consistent system of axioms* – an axiom system free from contradiction – in which such paradoxes would be avoided.

Russell's paradox resulted in Frege's eventual abandonment of his ambitious programme. Bertrand Russell himself, however, devoted much effort over several years to dealing with the paradox he had uncovered. He hoped to avoid antinomies by developing a complex 'theory of types', creating an elaborate hierarchy of different types of sets where at each level a set could only contain sets of lower types. Collaborating with *Alfred North Whitehead* (1861-1947), he produced the massive *Principia Mathematica*, published in several volumes from 1910 onward, in which they famously arrive at a proof of $1 + 1 = 2$ only after 379 pages.

In many ways the *Principia* represents the culmination of the logicist project to produce a complete set of axioms and rules of inference *within*

---

[15]In what is today called 'naive' set theory, this statement is the Comprehension Axiom, which asserts that for any (well-formed) formula $\phi(x)$ that contains x as a 'free' variable, we can obtain the set $\{x : \phi(x)\}$ whose members are precisely those objects that satisfy the proposition represented by ϕ. Examples (beloved of logicians) are 'x is a teacup' or 'x is a man'. But we can also include the empty set as defined by $\{x : x \neq x\}$, or, more precisely, we can define $\phi(x)$ via the statement $x = x$ and denote its negation by $\neg\phi$, so that $\varnothing = \{x : \neg\phi(x)\}$.

[16]In the notation of the previous footnote, Russell's set R is given by $\{x : \neg\phi(x)\}$, where ϕ represents the proposition $x \in x$.

The following popularised version of Russell's Paradox is well-known: In a certain village, there is a single barber. Every man in the village either shaves himself or he is shaved by the barber. Who shaves the barber? (Note the assumption that the barber is a man.)

symbolic logic from which, in principle, all mathematical truths would follow. It harks back to Leibniz' search for a *characteristica universalis*, a universal symbolic language in which concepts and ideas could be communicated effectively. But Russell never declared himself fully satisfied with his own efforts, and the focus of the debates about the meaning of mathematical statements shifted to debates about the specific *system of axioms* that would deliver a consistent theory of sets.

CHAPTER 10

Solid Foundations?

Democritus said: 'That truth did lie in profound pits, and when it was got it need much refining.'

Sir Francis Bacon, in *A Collection of Apophthegms, New and Old*, 1625

Summary

In this final chapter we observe how various paradoxes led to the effective abandonment of the logicist programme in favour of the establishment of an axiom system for set theory—initiated by Ernst Zermelo in 1908—that avoids these paradoxes and that has (to date) not been shown to be inconsistent. We then focus on debates about the Axiom of Choice, which was not included in Zermelo's system, but first made explicit in his proof of Cantor's Well-Ordering Principle. It created lively debates about permissible proof methods in mathematics, particularly in France.

Differences in perception between the two giants of mathematics around the turn of the twentieth century, Poincaré in France and Hilbert in Germany, later developed—via the injection of the *intuitionist* philosophy of the Dutch mathematician L.E.J. Brouwer into this debate—into open conflict and a crisis of confidence in the foundations of the subject. A technical discussion of these arguments is beyond the scope of this book and we present only a brief outline.

Hilbert's hopes of founding mathematics on a *formalist* viewpoint, avoiding all discussion of the 'nature' of mathematical objects, was dealt a lethal blow in 1930 by the *incompleteness theorems* of *Kurt Gödel*. Thus, advances in mathematical logic, a subject that owes its early development to the questions raised by the work of Cantor and Dedekind, ultimately forced mathematicians to adopt a more cautious attitude to the nature of mathematical truth.

I cannot hope to improve on the succinct, non-technical, description given to these events by John von Neumann, written in 1947 and quoted here at some length. The final section of this chapter, however, is devoted to a

https://doi.org/10.11647/OBP.0236.10

brief description of how *infinitesimals*, having been banished in the late nineteenth century, have made a modest comeback since 1960, notably through
the efforts of the logician Abraham Robinson.

1. Avoiding paradoxes: the ZF axioms

Although held in high regard by specialists, Russell and Whitehead's
Principia did not have a decisive influence on the directions of mathematical
research in the early decades of the twentieth century. In part, this was due
to the formulation, by Ernst Zermelo in 1908, of an apparently consistent
system of axioms for set theory *within mathematics*, using the notion of *set* as
an undefined term (much as Euclid does with *point* and *line*, despite appear-
ances). Zermelo's axiom system was later added to and completed by *Abraham Fraenkel* (1891-1965) and *Thoralf Skolem* (1881-1963) and is today known
simply as *ZF*. The relationship of the ZF axiom system to key questions that
Cantor's groundbreaking work had left unresolved, such as whether every
set can be well-ordered and the Continuum Hypothesis, became a significant topic of research over the next half-century.

Any precise specification of the axioms of *ZF* requires a background
in formal mathematical logic—which was *not* flagged as a pre-requisite for
reading his book! We must therefore be content with a brief informal description of the restrictions on set formation that Zermelo and his successors
considered necessary to avoid antinomies such as the ones indicated at the
end of the previous chapter.

Cantor had already expressed the hope that avoiding the use of notions
such as 'the set of all sets' of a specific kind might suffice to banish contradictions, but he never specified how this might be done without also inval-
idating proofs that appeared to produce useful correct results. Russell also
suggested early on that avoiding notions such as the *'class of all entities'* of
a particular type would lead *'naturally'* to the view that the set of objects
satisfying a functional proposition should be required to be equipotent to
some initial segment of the ordinal numbers. These ideas, however, did not
distinguish successfully between all types of antinomies.

In particular, a 'semantic' paradox first described by the French mathematics teacher *Jules Richard* (1862-1956) in 1905, applied Cantor's second
diagonal argument to the set E of all decimal expansions between 0 and 1
that can be *defined* (in English, say) *in a finite number of words*. This set must
be countable, and thus can be presented as a sequence $(r_n)_n$ of real numbers: start with all two-letter combinations in alphabetical order, then all
three-letter combinations similarly, and so on. Delete all those that do not
define a real number between 0 and 1 (represented here by an infinite decimal expansion). Thus the set E of all real numbers that can be defined in
a finite number of words can be written as a sequence, that is, as a well-
ordered denumerable set $\{r_1, r_2, ..., r_n, ...\}$. Now, by using this sequence (in

other words, by making reference to E), Richard describes how to define a decimal expansion (hence a real number) that is not equal to any of the r_n. Call this decimal expansion x. If the decimal expansion defining r_n has digit p in the nth place and p is neither 8 nor 9, then the nth digit of x will be defined as $p+1$. If the nth digit of r_n is 8 or 9 we take the nth digit of x as 1. But now we have used a finite number of (English) words to define x and x is not in E, which contradicts the definition of E.

Richard's paradox was picked up by *Henri Poincaré* (1854-1912), then the undisputed doyen of French mathematics, who had remained critical of various aspects of research into the foundations of mathematics. Poincaré argued that the paradox arose precisely because the collection E itself is used in defining x; in other words, we end up in a 'vicious circle'. According to Poincaré this could be avoided by defining E simply as *'the aggregate of all the numbers definable by a finite number of words without introducing the notion of the aggregate E itself'*. Peano and Zermelo disagreed, since a 'circular definition' is normally one that uses the term to be defined in the expression defining the term itself, and this was not the case here. But the paradox remained.

The use of terms such as *definable* itself leads to difficulties, such as in the *Berry* paradox, which Bertrand Russell ascribed to *GG Berry* (1867-1928), an Oxford librarian. Berry observed that the phrase *'the smallest positive integer not definable in under sixty letters'* (in the English language) is problematic. Only finitely many positive integers (i.e. natural numbers) can be described in under sixty letters: the alphabet has 26 letters, so for each letter we cannot have more than 26 choices. This means that the number of possible phrases (whether they make sense is irrelevant here) containing fewer than sixty letters is certainly *finite.* As there are infinitely many natural numbers, the set $U \subset \mathbb{N}$ whose members *cannot* be described in a phrase of under sixty letters is non-empty, hence by the well-ordering property of $\mathbb{N}$ it has a least member. On the other hand, if Berry's phrase describes a positive integer, it has done so with fewer than sixty letters! So the phrase is self-contradictory.

Zermelo's axiom system in 1908 sought to avoid both the antinomies described by Cantor and those involving 'definability', such as Richard's or Berry's. A restriction had to be placed on the ways in which we can generate subsets of a set by insisting that only certain kinds of logical sentences (sentential forms) could be used to identify the subset. Zermelo called such assertions *definite*; his description of this restriction was made fully precise by Hilbert's former student *Hermann Weyl* (1885-1955) in 1910, who specified the logical symbols that could be allowed.

Unlike Euclid's five axioms for geometry (with all their shortcomings and omissions, which were only rectified by Hilbert in 1899), any summary statement of the ZF axioms will not produce much enlightenment in the reader who encounters them for the first time. (Nevertheless, a summary is attempted in *MM.*) Suffice it to say for our purposes that they were designed explicitly to avoid the pitfalls that were described earlier. They provide a *procedure* for set formation which, to date, has not been shown to lead to inconsistencies.

It was also gradually becoming clear that mathematics, if based upon axioms in this fashion, could not be subsumed under logic, as Frege, Russell and others had hoped.[1] For their programme to succeed, the logicists, led by Frege's meticulous work, had extended classical (Aristotelian) logic to include logical symbols and quantifiers which enabled them to identify more precisely what constitutes a *logical sentence* or *proposition P*. Such a sentence should be expressible in terms of the symbols comprising the given logical language, and be derived from logical axioms, independently of any meaning that one might attach to the symbols that make up the proposition. The ZF axiom system, on the other hand, includes theory-specific assumptions, such as the Axiom of Infinity, that do not adhere fully to these constraints.[2] Its success in avoiding the antinomies discussed earlier and its subsequent acceptance by the mathematical community has meant that logicist attempts to subsume all of mathematics under logic were gradually abandoned.

2. The axiom of choice

Instead of discussing the Zermelo-Fraenkel axioms individually, we consider an axiom not included in his list by Zermelo, but used explicitly by him in an earlier important paper in 1904. In this paper he considered Cantor's contention that *every set can be well-ordered*. Zermelo wished to justify Cantor's claim, which would serve to complete the foundations of Cantor's theory of transfinite numbers. Recall that Cantor had managed to prove that the trichotomy holds for well-ordered sets (such as any two alephs), but that a proof of this had eluded him if the sets were not well-ordered. Proving that any power is an aleph would require him to show that any set can be well-ordered, and this was an open problem until Zermelo's paper. However, in presenting a proof of Cantor's contention, Zermelo stated and used an unproven assumption (he called it an *'unobjectionable logical principle'*) that soon became known as the *Axiom of Choice*. This assumption was to play a fundamental role in much of modern mathematics.

2.1. Initial reception. Zermelo's statement of the assumption in his original 1904 paper can be formulated more succinctly in the following form:

---

[1]Readers looking for a more detailed account of the issues we touch upon in this chapter may consult the excellent article [43].

[2]The axiom allowing us to go beyond the finite, *Infinity*, postulates the existence of an infinite set. More specifically, the axiom states that there is a set Z containing the empty set $\varnothing$, and for any A in Z, the union $\cup\{A, \{A\}\}$ is also in Z. The infinite set then arises from the indefinite repetition of this operation in a manner similar to von Neumann's model of $\mathbb{N}$, discussed in **Chapter 8**.

Given any family T of non-empty sets, there is a function f which assigns to each member A in T an element $f(A)$ of A.

The function f is called a *choice function*. So the Axiom of Choice says that a choice function always exists for any family of non-empty sets.

In 1908 he reformulated his axiom slightly, to assert the existence of a choice set, or *transversal*, for any collection of sets: *For every collection A of mutually exclusive non-empty sets there exists at least one set containing exactly one element from each member of A.*

In this formulation it is easier to appreciate the analogy used by Russell (slightly tongue-in-cheek, as was typical of him) to explain why this axiom has any content. The claim is clearly true for any finite collection of sets. Russell points out that the axiom demands that the formation of a set consisting of one element from each set in the given collection should always be possible, whether or not we can specify a 'selection rule' for doing so. To illustrate this, he imagines two infinite collections: the first consists of pairs of shoes, the second of pairs of socks. A obvious selection rule for the shoes would be to choose the left shoe of each pair, but for socks it is quite unclear how one might define a selection rule, since in any given pair the left and right socks look identical!

Note also that, while the Axiom asserts the existence of a choice function for any family T of sets, the 1908 formulation can easily be amended in various ways to an apparently weaker claim. For example, one might restrict the assertion to cases where T is a set whose cardinality is at most α for some specified α. The weakest of these is *denumerable choice*, where one would only claim the existence of a transversal for every countably infinite set T. As we will see below, many nineteenth century researchers in Analysis had implicitly made this assumption. By contrast, Zermelo needed the full power of the Axiom, applied to an arbitrary family T of sets, in his efforts to prove Cantor's Well-Ordering Principle.

Despite its innocuous appearance, the Axiom of Choice has aroused more controversy than any axiom in the history of mathematics, with the possible exception of Euclid's Parallel Postulate. David Hilbert called it the axiom *'most attacked up to the present in the mathematical literature'*. Although Kronecker, who might have been its most vehement critic, had died more than a decade earlier, other figures took up the cudgels on his behalf.

In particular, several prominent younger French mathematicians of the time, including *Emile Borel* (1871-1956), *Henri Lebesgue* (1875-1941) and *René-Louis Baire* (1874-1932), expressed grave doubts about the validity of the assumption in a now famous correspondence with *Jaques Hadamard* (1865-1963), who argued strongly for the acceptance of the axiom.

Borel had begun the debate with a short article in 1904, in which he argued that Zermelo's proof had simply shown the equivalence of two problems:

(A) whether an arbitrary set M can be well-ordered,

(B) whether it is possible to choose a distinguished element from each non-empty subset of M.

This equivalence, he maintained, did not amount to a solution of problem (A), since the problem of *determining* a distinguished element from an arbitrary subset of M seemed to him *'one of the most difficult, if one supposes, for the sake of definiteness, that M coincides with the continuum'*. The acceptance of uncountably many arbitrary choices took one *'outside mathematics'*, Borel claimed. He was strongly supported by Baire, who took matters even further, since he rejected the actual infinite (such as Cantor's transfinite ordinals) and argued that it is false, when considering an infinite set, *'to regard the subsets of this set to be given'*.

On joining the debate at Borel's request, Lebesgue initially took a more cautious approach. For him the key question was whether one could *prove the existence of a mathematical object without defining it*. He concluded that defining the object uniquely was essential – even though, in his own work, he had at times used existence proofs that did not conform to this requirement. For him, Zermelo's use of an infinite number of arbitrary choices could not have meaning as an existence proof. Lebesgue went further than Borel in rejecting the possibility of Denumerable Choice, and the proposition that any infinite set has a denumerable subset.

Hadamard, in reply, argued that the central problem was that he and his colleagues had different conceptions of mathematics, and that their arguments resembled earlier debates around Riemann's views on what functions should be allowed into analysis. In his view, *'essential progress in mathematics has resulted from successively annexing notions'* which, for earlier generations, *'were "outside mathematics" because it was impossible to define them'*.

This correspondence (here taken from [**32**]) was published in a major French mathematical journal in 1905, and represents an early instance of formation of a French *constructivist* school of mathematics. In time, Baire, Borel, and especially Lebesgue – all of whom (like the majority of late nineteenth century analysts) had implicitly used instances of the Axiom in their own work – became strong protagonists for the conception of mathematics they had spelled out in their initial responses to Zermelo's Axiom. Lebesgue even called their approach 'Kroneckerian'.

2.2. Earlier uses of the Axiom. The various ways in which late nineteenth century mathematicians had used the Axiom of Choice prior to Zermelo's formal statement of the assumption are analysed comprehensively in

[32]. The example we consider concerns the crucial concept of *continuity of a function* $f : \mathbb{R} \to \mathbb{R}$ *at a point* a (cf. **Chapter 6**).

Recall that Bolzano's definition of continuity was rephrased there to read

(i) *The function f is continuous at the point a if, for given $\varepsilon > 0$ we can find $\delta > 0$ such that $|f(x) - f(a)| < \varepsilon$ whenever $|x - a| < \delta$;*

wheras Cauchy's definition lent itself to the reformulation

(ii) *The function f is continuous at the point a if whenever a sequence (x_n) has limit a, the sequence of values $(f(x_n))_n$ has limit $f(a)$.*

As noted in **Chapter 6**, (i) states the modern definition of continuity, while the formulation (ii) is now called *sequential continuity.* For real functions (though not in more general settings) (i) and (ii) are logically equivalent.

However, the proof that sequential continuity implies continuity implicitly uses Denumerable Choice. The following argument mirrors a proof given in an 1871 paper by Heine, who credited it to Cantor.

The argument goes as follows: if (i) fails for f at a, there must exist an $\varepsilon > 0$ such that for any $\delta > 0$ there is an x satisfying $x - \delta < a < x + \delta$, for which $|f(x) - f(a)| \geq \varepsilon$. Choose δ successively as the numbers $\delta_n = \frac{1}{2^n}$ to produce points x_n with $0 < |x_n - a| < \frac{1}{2^n}$, but $|f(x_n) - f(a)| \geq \varepsilon$. So: $\lim_{n\to\infty} x_n = a$, but $f(a) \neq \lim_{n\to\infty} f(x_n)$. This means that f does not satisfy (ii) at a. Therefore: if (ii) is true for f at a then (i) must also be true for f at a.

But how exactly do we describe a *rule* for choosing each of the points x_n? What has been shown is that the definition of continuity implies that if the function f is *not* continuous at a, then, for each $n \geq 1$, there must be a point x_n satisfying $0 < |x_n - a| < \frac{1}{2^n}$, but whose images under f are at least ε apart. However, we have no explicit way of calculating the value of the number x_n. Even a rule such as taking x_n as the 'smallest' such point does not provide a way of *identifying* the point x_n explicitly.

This is an example where the Axiom of Choice (at least in its Denumerable guise) cannot be avoided. This fact was only noticed in 1913 in Italy by *Michele Cipolla* (1880-1947), and picked up in Poland in 1916 by the influential *Wacław Sierpiński* (1882-1969). The latter founded a group of mathematicians in Warsaw who embarked upon an exhaustive study of the Axiom of Choice, beginning with an extensive survey by Sierpiński of uses of the Axiom in Real Analysis. In addition to the example given by Cipolla, he described numerous key results by Borel and Lebesgue that had made implicit use of the Axiom. One striking example was that the proof of the principal characteristic of Lebesgue measure (see [3]), *countable additivity*, in fact depended on Denumerable Choice, whose validity Lebesgue later explicitly rejected in the correspondence summarised above!

Over the following years the Axiom of Choice became a touchstone for researchers on the foundations of mathematics. The Axiom perhaps represents the main element of what remains of Cantor's bold claim of the *'freedom'* inherent in pure mathematics as a creation of the human spirit that does not need to be kept within artifical boundaries—such as those insisted upon by Kronecker.

Such restrictions, Cantor had argued, represent a far greater danger to mathematics than did his own firm belief, that *'mathematics is completely free in its development'*, and bound only by the requirement that its concepts are *'free from contradictions in themselves'* as well as *'standing in fixed relationships, ordered through definitions, to earlier concepts that are already present and have been verified'*. Even Poincaré, who argued consistently for the primacy of *intuition*, declared himself disposed to accept the Axiom. Ever the Kantian, he declared the Axiom of Choice to be *'a synthetic **a priori** judgment without which the "theory of cardinals" would be impossible, for finite as well as infinite numbers'*.

Most mathematical practitioners today accept the ZF axioms as the foundation on which mathematical concepts can be built, beginning with set theory and specialising to whatever their field of interest may be. Most also use the Axiom of Choice as a valuable tool—and we will see below that, if one accepts the ZF axioms, adding the Axiom of Choice as an additional axiom creates no new logical difficulties. When this is done, the resulting axiom system is simply called ZFC.

3. Tribal conflict

In the early years of the twentieth century, foundational disputes arose between various schools of mathematicians, in part over the use of Axiom of Choice and partly over more profound philosophical and methodological differences. While these are issues of fundamental importance to the subject as a whole, they have, in large measure, tended to become the province of specialists in the foundations of the subject.

3.1. Hilbert and Poincaré. The contrasting philosophical perceptions of Henri Poincaré in Paris and David Hilbert in Göttingen were already apparent in their reactions to Kronecker's attacks on Cantor's work. Hilbert had admired Cantor's transfinite mathematics from the beginning, without allowing himself (at least initially) to become too disturbed by the antinomies of set theory. In 1925, more than three decades after Kronecker's death, Hilbert delivered an address entitled *On the Infinite* in Münster at an event honouring Weierstrass' work in creating more solid foundations for Real

Figure 40. David Hilbert, by an unknown photographer, 1907[3]

Analysis, He again hailed Cantor's set theory as *'the finest product of mathematical genius and one of the supreme achievements of purely intellectual human activity'* (see [**37**]).

However, Hilbert was fully aware of the damage the antinomies would do unless ways were found to avoid them. In contrast to Kronecker's 'arithmeticisation programme', his proposal was to *'supplement the finitary statements'* of ordinary arithmetic with *'ideal statements'*, just as ideal elements had been used in parts of function theory and geometry in the past. Mathematics would then consist of two kinds of statements: those to which meaning could be attached by some external communication, and others which had no meaning in themselves, but were the ideal elements of the theory. The sole criterion for their validity would be a *'proof of consistency'*, arrived at via a (formal) logical calculus. In this way, Hilbert declared triumphantly, *'No one will drive us from this paradise that Cantor has created for us!'*

Today, Hilbert's approach to foundational questions is described as *formalism*, which is summarised in *Encyclopaedia Britannica* as the belief that *'all mathematics can be reduced to rules for manipulating formulas without any reference to the meanings of the formulas. Formalists contend that it is the mathematical symbols themselves, and not any meaning that might be ascribed to them, that are the basic objects of mathematical thought'*.

---

[3]https://commons.wikimedia.org/wiki/File:David_Hilbert,_1907.jpg

The position of most mathematicians is somewhat different. Recalling Plato's insistence on the reality of his World of Ideas, one might argue that a typical mathematician today will act as an unreconstructed *Platonist* on weekdays, pausing only to recite *Formalist* liturgy when required to do so on particular feast days.

Hilbert's purpose, following on from his successful axiomatisation of Euclidean geometry, was to find a general methodology (which he called *proof theory*) for constructing a formal language in which the *consistency* of the axiom system in question could be verified. He had shown in 1899 that Euclidean geometry is free from contradictions *provided* one could assume that arithmetic and the real number system (the basis of Analysis) had this property. But this result, rather than solving the problem of consistency, had shifted it to another area of mathematics. The real goal was to obtain a proof of *absolute consistency* of certain kinds of formal systems.

This had brought the focus onto the question of the consistency of arithmetic (the system codified by the Peano axioms); in other words, formulating this system in the corresponding formal language, and using only the axioms of ZF and those of classical logic to provide a formal proof (within the formal language) that it does not contain a contradiction. The essence of the *Hilbert programme* was to construct mathematical proofs that the various branches of the subject are free from contradiction.

Poincaré, by contrast, would always emphasise the importance of what he called *intuition* in mathematical research. While praising Hilbert's *Foundations of Geometry* as a 'classic', he remarked that *'the logical point of view alone appears to interest Professor Hilbert... ...The axioms are postulated; we do not know from whence they come...'* Defending Kant's philosophical concept of *a priori* knowledge, Poincaré argued that it was not inherent notions of space or time, but rather the concept of *iteration*, or indefinite repetition (as evidenced in counting), which required our innate sense of time. This constituted the *a priori* source of extra-logical content in elementary number theory. Moreover, he argued that an intuitive grasp of 'continuity' is basic to our understanding of the continuum. It was such *a priori* precepts, rather than logical reasoning, that enabled us to understand the underlying basis of mathematical knowledge.

He rejected Cantor's views on the actual infinite, which he saw as leading to a vicious circle of self-referential statements, as illustrated by the various antinomies. His philosophy also moved him strongly to oppose the logicist programme: he did not accept that mathematics could be a mere part of logic, leaving no room for intuition as an innate characteristic of the intellect. In his view, Hilbert's relegation of truth in mathematics to mean simply the absence of contradiction, and his proposed proof theory in particular (which existed only in merest outline until after Poincaré's early death in 1912), were in danger of straying too close to the logicist perspective.

3.2. Enter Brouwer. A more substantive challenge to 'classical' mathematics and logic was developed between 1908 and 1918 by the Dutch mathematician *L.E.J. Brouwer* (1881-1966), whose incisive contributions to the burgeoning subject of topology had marked him out as a brilliant researcher. The publication of his ideas on the nature and foundations of mathematics began with a paper published in 1918, entitled *Founding Set Theory Independently of the Principle of the Excluded Middle. Part One, General Set Theory.*

Brouwer's stated objective was to remove the use of proof by contradiction from mathematical arguments dealing with infinite sets. He objected to the unrestricted use of the Excluded Middle as a logical tool to claim the *existence* of a mathematical object, reflecting the concerns of the French constructivists (Borel, Baire, Lebesgue, as well as Poincaré), that proof by contradiction would not identify the object in question uniquely.

Brouwer produced a number of simple, but compelling, examples to draw attention to his claim that the construction of the classical models of the continuum cannot always *determine* particular real numbers. For example, suppose that the expansion of the real number x starts with 0.33333... and then *either* continues in this fashion forever, *or* is terminated as soon as a string of seven consecutive sevens (....7777777...) appears in the decimal expansion of π.

In Brouwer's time no such string had been found in the decimal expansion of π, so it was not known if one existed. (If one were to be found, he could always demand a longer string, of course.) But a string of sevens must either appear at some point, or else it never appears. This means that the number x cannot be irrational: it equals $\frac{1}{3}$ if the string never appears, while its decimal expansion terminates at the end of the string if it does appear. But we cannot *identify* the rational number in question, since we don't know whether or at which decimal place the expansion terminates. Similar examples, usually based on an unsolved number theory problem (see *MM* for an example) were constructed to exhibit a real number x for which the question whether $x = 0$ or $x \neq 0$ required the solution of the unsolved problem. Brouwer regarded the very existence of unsolved problems in mathematics as a *weak counterexample* to the Law of the Excluded Middle.

Thus, for Brouwer, the key requirement was that only sets that could be constructed were to be accorded meaning in his *intuitionist* mathematics. Everything should spring, ultimately, from our fundamental intuition of the sequential construction of the natural numbers. He was uncompromising: rather than claim that the results of classical mathematics were wrong, he argued that very many of them should be regarded as meaningless!

Moreover, it was clear that even those results that could be retained as meaningful in intuitionist terms would usually require much longer and less elegant proofs if the available mathematical tools were stripped down to those deemed acceptable by Brouwer. Hilbert memorably described the requirements of intuitionist methods as akin to *'denying the boxer the use of his fists'* when attacking a particular problem.

Throughout the 1920s, Brouwer's trenchant critique of 'classical' mathematics and logic attracted a dedicated following of talented mathematicians—mainly in Holland and Germany—who were persuaded by his arguments of the need for change. Notably, he increasingly influenced Hilbert's favourite former student, *Hermann Weyl* (1885-1955), whose enthusiastic conversion to intuitionism would cause Hilbert much personal consternation. In 1918 Weyl, a highly skilled communicator, published his own version of semi-intuitionist analysis, *The Continuum*. This became influential in Germany, as did his 1921 paper on the 'new foundational crisis' in mathematics.

The debates between Weyl, Brouwer and Hilbert were courteous enough at the outset, but later the dispute between Brouwer and the normally generous and fair-minded Hilbert took on a personal tone. In 1928, in an episode reminiscent of the battle between Kronecker and Cantor, Hilbert (who was seriously ill at the time and did not expect to survive) used his pre-eminence among German mathematicians to engineer the removal of Brouwer from the editorial board of the prestigious journal *Mathematische Annalen*. This led to considerable ill-feeling and Albert Einstein resigned from the board in protest. Ironically, this time the senior participant in the battle was the adherent of Cantor's 'free' mathematical style, while the outcast and principal sufferer was the man who had sought to 'purify' mathematics from the effects of that style and return it to safer foundations.

Today there are various strands of intuitionism within the somewhat wider (but still quite limited) group of *constructivist* mathematicians, among whose notable achievements the work begun by the founder of modern constructive analysis, the American *Errett Bishop* (1928-1983), perhaps stands out as having attracted most interest. The great majority of mathematicians, however, continues to work within 'classical' mathematics.

4. Gödel's incompleteness theorems

The final act in the 'foundational crisis' occurred only three years later. In 1931 the Austrian logician *Kurt Gödel* (1906-1978) dealt a fatal blow to the Hilbert programme by publishing his two (now famous) *incompleteness theorems*. An understanding of what these theorems said and why they had such a devastating impact upon the formalist ideal necessitates a more detailed description of that ideal.

4.1. Hilbert's programme. In Hilbert's view, the provision of an axiomatic basis for a branch of mathematics—as he had done for Euclidean geometry in 1899—ensured that one could ignore the traditional 'meaning' of the concepts encountered in the theory: as he often emphasised, notions such as *point, line* or *plane* should be capable of being replaced by *tables,* *chairs*, or *beer mugs*! They would become undefined terms, whose properties as well as the relations between them would be determined by using only the stated axioms. In the *formalisation* of a mathematical theory its objects appear as 'meaningless signs' expressed in a very precise, formal logical language whose syntax (or grammar) expresses the rules for combining individual signs into longer strings, or 'sentences' of the language.

In such a formalisation the proof of a mathematical theorem comprises a series of valid steps—derived from the axioms specific to that area of mathematics and expressed in terms of the symbolic language, while obeying the logical axioms and rules of of inference—that transform one such sentence into another. In this way one can ensure that no hidden assumptions or unstated logical principles have crept into the deductive process. The relationships between the different sentences representing mathematical statements are demonstrated explicitly and completely

It then becomes possible to ask specific questions about the nature of and relationships between different strings of symbols. Among these is to ask whether it is possible, within the system, to construct a sentence, say ϕ, such that ϕ and its negation $\neg\phi$ are both consequences of the underlying set of axioms. The sentence ϕ might, for example, be expressed within the formal language through the formula '$0 = 0$'. Its negation would be expressed as the formula '$0 \neq 0$'. The question is whether both can be deduced from the underlying axiom system.

A proof that such a situation cannot occur within the system would verify that the formalised system is *absolutely consistent*, that is, free from inherent contradiction. The objective of Hilbert's program was to achieve this aim for each area of mathematics by what he called *finitary reasoning*, and so rebut a key element of Brouwer's critique of classical methods.

To illustrate Hilbert's approach we consider the case of *arithmetic*, as codified by the Peano axioms. To express these precisely requires a suitable and well-defined logical language. In mathematical logic, a *first-order language L* for an axiom system S would satisfy five requirements, only one of which is specific to S. The general requirements for L are a (denumerable) list of *variables* $(x, y, z, ...)$ and three types of logical symbols: *connectives* (symbols for 'not' ($\neg$), 'and '($\vee$), 'or' ($\wedge$) and 'implies' ($\rightarrow$)), *equality* (=) and the *quantifiers* 'for all' ($\forall$) and 'there exist' ($\exists$). The symbols specific to S are its *undefined terms*: For Peano arithmetic these are the notions of 'distinguished' element (0), 'immediate successor' (s), and the operations of addition (+) and multiplication ($\times$).

The Peano axioms (expressed in this formal language), together with logical axioms and rules of inference, are used to derive the *theorems* of S, i.e. the statements that can be proved using just the axioms and rules of inference. No meaning is attached to these statements—as noted in [33], a page 'of "meaningless marks" of such a formalised mathematics does not **assert** anything'.

Hilbert's programme, as well as his conception of what constituted *finitary reasoning*, were to evolve substantially throughout the 1920s.[4] His goal, however, remained to find, for any mathematical theory, a way of proving, entirely *within* the formal language, that it is impossible to arrive at a contradiction when using only the (finitely many) given axioms specific to the theory as well as the logical axioms.

If this could be achieved, Hilbert argued, he would finally have laid the ghost of Kronecker to rest. He would also have demonstrated that, by accepting consistency as the standard of truth for a mathematical system, its theorems could be seen to have solid foundations and thus could safely be used and explored further, without subjecting their proofs to Brouwer's more demanding criteria.

In its grand ambition, as well as in its careful attention to detail, Hilbert's programme is perhaps reminiscent of Frege's and Russell's attempts–in their different ways—to subsume mathematics under formal logic. And while *Principia* provided most of the new tools needed for Hilbert's formalisation of arithmetic and the rules of inference in logic, it was the methodology of this process which laid the groundwork for Gödel's investigations. The question whether there is a proof of consistency for a given axiom system is a statement that is not phrased *within* the logical language constructed to that system – it is a statement *about* the system itself, a *meta-mathematical* question.

4.2. Decidability and consistency. This is reminiscent of the difficulty that underlies *Richard's paradox,* which we discussed at the beginning of this chapter. In fact, Gödel confirmed that his arguments were to some extent modelled on reasoning suggested by this paradox, namely that its statement *appeared* to be about decimal expansions, but was actually concerned with the properties of the set E of real numbers definable in a finite number of words. By listing the various combinations of letters one can attach a number to each such combination, representing its position in the sequence.

---

[4]A detailed description of the philosophical implications of Hilbert's evolving use of the term *finitary* is given in the Stanford were to evolve throughout the 1920s, were to evolve throughout the 1920s, of Philosophy under the heading 'Hilbert's Program' – the details are too technical to include here. In [43] a definition of 'finitary', given by the French formalist *Jacques Herbrandt*, is reproduced as follows: 'By a finitary argument we understand an argument satisfying the following conditions: In it we never consider anything but a given finite number of objects and of functions; these functions are well defined, their definition allowing the computation of their values in a univocal way; we never state that an object exists without giving the means of constructing it; we never consider the totality of all the objects x of an infinite collection; and when we say that an argument (or a theorem) is true for all these x, we mean that, for each x taken by itself, it is possible to repeat the general argument in question, which should be considered to be merely the prototype of these particular arguments.'

We apply the same idea to all *definitions* that define a property of numbers (e.g 'k is prime'). The list of these definitions (expressible in English, say) is countable, so they may be written as $(\phi_k)_{k \geq 1}$, for example. We can check whether any given number has the property expressed by ϕ_k (e.g if $\phi_k(n)$ expresses 'n is prime', then $\phi_k(71)$ holds, and so does $\neg\phi_k(65)$, since 71 is prime, while 65 is not). The formal equivalent of the statement 'k *does not have the property designated by the defining expression with which k is correlated in the serially ordered list of definitions*' then becomes $\neg\phi_k(k)$. Following [33], let us call such a number *Richardian*. But now 'n is Richardian' is also a definition in the list, hence it is allocated a number, say m. Then the question: *is m Richardian?* is self-contradictory: we now have a natural number m such that for all k, $\phi_m(k)$ holds if and only if $\neg\phi_k(k)$ holds, and therefore $\phi_m(m)$ holds if and only if $\neg\phi_m(m)$ holds.

Working in the formal system S representing Peano arithmetic, Gödel (using an adapted version of the system developed in *Principia*) was similarly able to 'mirror' certain meta-mathematical statements by statements *within* the language of S. By assigning a unique *Gödel number* to each elementary sign in the formal language, to each formula and to each proof (which is a finite sequence of formulas), he was able to analyse the structure of the formal language with great precision and formulate 'mirror images' *within* the language of S of various statements *about* the language—a fairly non-technical acount can be found in [33].

In his *First Incompleteness Theorem*, Gödel showed that, *if S is consistent, there is a statement G (a Gödel statement) in this language such that neither G nor its negation $\neg$G (not-G) can be proved within S.*

In other words, the theorem shows that S is an *incomplete* system.

Gödel's encoding of all statements and proofs in S by natural numbers (expressed as products of powers of their prime factors in ascending order) enabled him to assign a Gödel number g to the statement G, expressed in the language of S, that mirrors the (meta-mathematical) statement '*This statement is not provable*'.[5] The statement G is codified by its Gödel number g, so we can read this claim as '*The statement with Gödel number g is not provable*'. Gödel next showed that if G were provable in S, then the same would hold for its negation $\neg G$, which would mean that S is inconsistent, contradicting the hypothesis of the theorem.

The first theorem therefore says that the statement G is (formally) *undecidable in S*.[6]

---

[5]Note the close similarity to the *'who shaves the barber?'* version of Russell's paradox (see Footnote 17 in **Chapter 9**) or the so-called *Liar paradox*, stating *'This statement is false'*.

[6]However, the statement G is true. It makes the (meta-mathematical) assertion: 'there is no proof of G within S', which is a true statement, since we have just seen that G is undecidable in S. Thus: provability of a statement within a formal system and truth are not the same thing!

Adding either G or its negation as a new axiom to S will not improve matters. Gödel's numbering technique works for any axiom system that (like S) is rich enough for arithmetic while remaining *computable*—so that its axioms can be recognised by a computer. Adding 'Gödel statements' to the axioms one by one does not destroy this property. In other words, S is *essentially incomplete*.

Gödel's *Second Incompleteness Theorem* followed from the first. Informally, it states:

If S is consistent, then that fact cannot be proved within S.

To justify this claim, Gödel first codes the statement 'S is consistent' by $Consis(S)$, while the statement 'G is not provable' is coded by G itself. The First Incompleteness Theorem states that if S is consistent then G is not provable in S. Coding this statement in S as $Consis(S) \to G$, the proof of the first theorem can be mirrored in S to show that $Consis(S) \to G$ is provable in S, contradicting the first theorem. Therefore the consistency of S *cannot* be proved within S.

This conclusion was a hammer blow to Hilbert, showing that the original objective of his proof theory programme is unachievable. If there were a finitary demonstration of the consistency of S, it should be possible to formulate it as a theorem of S, which would mean, in turn, that $Consis(S)$, and therefore G, would be provable in S, contradicting the first incompleteness theorem!

While this does not exclude the possibility of finding a 'finitary' proof (in Hilbert's sense) that *cannot* be mirrored in the formal language of *Principia* (which is essentially the system we called S) but which shows the consistency of arithmetic, no such proof has yet been found. Work by various of Hilbert's former co-workers and others has produced partial results, but his original goal has effectively been abandoned as hopeless.[7]

Thus the optimistic claim that '*every mathematical problem can be solved*' and Hilbert's stirring epitaph: '*Wir müssen wissen. Wir werden wissen.*' ('We must know. We will know.') have, for the present at least, had to be replaced by somewhat more modest objectives. Gödel's results are a salutary reminder that the axiomatic method itself – at least as understood today – has severe limitations.

In 1938/9 Gödel lectured in the USA, but, showing little interest in politics, he returned to Vienna despite Nazi Germany's takeover of Austria in 1938. He did not succeed in obtaining a paid position—possibly because he had many Jewish friends. He was even mistaken as Jewish and attacked on

---

[7]In 1936, *Gerhard Gentzen* proved the consistency of S, imposing a linear ordering on statements in S. However, this went far beyond finitary statements: he needed not only the ordinals of Cantor's class (I), but ordinals up to a certain *infinite* ordinal (ε_0).

the street. He obtained a US visa in 1940, having to travel to the US via Russia and Japan, and settled at the Institute for Advanced Study in Princeton, where he remained until his death in 1978—and where he formed a close friendship with *Albert Einstein*.

Soon after arriving in Princeton he published *Consistency of the axiom of choice and of the generalized continuum-hypothesis with the axioms of set theory*, now regarded as a classic. He defined, within ZF, a smaller collection of sets (which he called *constructible*) in which the axioms of ZF, as well as the Axiom of Choice (AC) and the Generalised Continuum Hypothesis (GCH), are true. Hence, if the *negation* of AC could be proved in ZF, that would probably also hold within the constructible universe, and would make ZF inconsistent.[8] The same would apply with GCH. So: if ZF is consistent, then so is the axiom system ZF+AC+GCH—no *new* inconsistencies can arise by adding AC and/or GCH to ZF.

In 1963, *Paul Cohen* (1934-2007), used a revolutionary technique (*forcing*) to show that, provided ZF is consistent, it also remains so if we add the *negations* of AC and/or GCH to it instead. Hence, if ZF is consistent, then neither AC nor GCH can be *proved* within ZF. Combining this with Gödel's result, therefore, AC and GCH are *independent* of ZF, i.e. neither provable nor disprovable.

In this more modest sense, and despite effectively ending Hilbert's hopes of proving absolute consistency, Gödel was instrumental in rescuing aspects of 'Cantor's paradise' that Hilbert had sought to preserve, as quoted at the beginning of this section.

4.3. von Neumann's verdict. John von Neumann summarised the impact of intuitionism and Gödel's theorems as follows in his essay *'The Mathematician'*, published in *'The Works of the Mind'* (vol.1, pp. 180-196, University of Chicago Press, 1947):

It is difficult to overestimate the significance of these events. In the third decade of the twentieth century two mathematicians—both of them of the first magnitude, and as deeply and fully conscious of what mathematics is, or is for, or is about, as anybody could be—actually proposed that the concept of mathematical rigour, of what constitutes an exact proof, should be changed! The developments which followed are equally worth noting.

1. Only very few mathematicians were willing to accept the new, exigent standards for their own daily use. Very many, however, admitted that Weyl and Brouwer were prima facie right, but they themselves continued to trespass, that is, to do their own mathematics in the old, "easy" fashion—probably in the hope that somebody else, at some other time, might find the answer to the intuitionistic critique and thereby justify them a posteriori.

---

[8]For a discussion of what is meant by *true* in this context, see article on 'The Axiom of Choice' in the *Stanford Encyclopedia of Philosophy*, http://www.plato.stanford.edu.

2. Hilbert came forward with the following ingenious idea to justify "classical"'
(i.e. pre-intuitionistic) mathematics: Even in the intuitionistic system it is possible
to give a rigorous account of how classical mathematics operates, that is, one can
describe how the classical system works, although one cannot justify its workings.
It might therefore be possible to demonstrate intuitionistically that classical proce-
dures can never lead into contradictions—into conflicts with each other. It was clear
that such a proof would be very difficult, but there were certain indications how it
might be attempted. Had this scheme worked, it would have provided a most remark-
able justification of classical mathematics on the basis of the opposing intuitionistic
system itself! At least, this interpretation would have been legitimate in a system of
the philosophy of mathematics which most mathematicians were willing to accept.

3. After about a decade of attempts to carry out this program, Gödel produced
a most remarkable result. This result cannot be stated absolutely precisely without
several clauses and caveats which are too technical to be formulated here. Its es-
sential import, however, was this: If a system of mathematics does not lead into
contradiction, then this fact cannot be demonstrated with the procedures of that
system. Gödel's proof satisfied the strictest criterion of mathematical rigour—the
intuitionistic one. Its influence on Hilbert's program is somewhat controversial,
for reasons which again are too technical for this occasion. My personal opinion,
which is shared by many others, is, that Gödel has shown that Hilbert's program is
essentially hopeless.

4. The main hope of a justification of classical mathematics—in the sense of
Hilbert or of Brouwer and Weyl—being gone, most mathematicians decided to use
that system anyway. After all, classical mathematics was producing results which
were both elegant and useful, and, even though one could never again be absolutely
certain of its reliability, it stood on at least as sound a foundation as, for example,
the existence of the electron. Hence, if one was willing to accept the sciences, one
might as well accept the classical system of mathematics. Such views turned out to
be acceptable even to some of the original protagonists of the intuitionistic system.
At present the controversy about the "foundations" is certainly not closed, but it
seems most unlikely that the classical system should be abandoned by any but a
small minority.

I have told the story of this controversy in such detail, because I think that it
constitutes the best caution against taking the immovable rigour of mathematics too
much for granted. This happened in our own lifetime, and I know myself how hu-
miliatingly easily my own views regarding the absolute mathematical truth changed
during this episode, and how they changed three times in succession!

Even while the foundational debates were raging in the 1910s and 1920s,
the great majority of mathematicians paid only occasional attention to these
matters, and got on with their particular research projects, while paying due

respect to the efforts of those grappling with difficult foundational questions. Following the shock of Gödel's 'negative' results, this tendency became more marked, while the pace of progress in many areas of mathematics increased dramatically, with new areas developing so quickly that no single researcher after Poincaré and Hilbert could truly be regarded as a 'universalist' with contributions and interest stretching across the entire spectrum of the subject. The emergence of computing has shifted some attention back towards the discrete and to the rapid development of recursive techniques as well as of rules of inference, including probabilistic techniques and multi-valued logics, but their impact on the fundamental questions around the nature of 'mathematical truth' is by no means clear at present.

5. A logician's revenge?

While the controversy over Cantor's transfinite cardinals raged in the period 1880-1900, there was nevertheless near-unanimity among mathematicians that, following the construction of arithmetical models of the continuum $\mathbb{R}$, the 'infinitely small' could no longer lay any claim to legitimacy in the foundations of the Calculus. Among the few notable dissenting views, those expressed by *Paul du Bois-Reymond* (1831-1889), especially in his *Die Allgemeine Funktionentheorie* (General Function Theory), were more strikingly and elaborately articulated than most (see [2]).

This work mixes a certain amount of mysticism and metaphysics with mathematical analysis in building contrasting pictures of the continuum that foreshadow later debates. In a dialogue between an 'idealist' and an 'empiricist', various conceptions of the continuum are analysed, but no final conclusions are reached. However, in an earlier paper du Bois-Reymond had developed a theory of infinitesimals, arguing that if infinite numbers were to be regarded as legitimate, so should their inverses, the infinitesimals.

Although du Bois-Reymond was not regarded as a mathematician of the first rank, he later made claims of priority in the use of the second diagonal method, which is universally attributed to Cantor. His more famous brother *Emil*, a physiologist who later became a philosopher, was somewhat more influential in late nineteenth-century society. Emil's writings focused on limits to our knowledge of nature, arguing that there were 'transcendental' questions that were unsolvable in principle. His statement *'Ignoramus et ignorabimus'* ('we are ignorant and we shall remain ignorant') became something of a rallying cry among groups of university students at the time. In 1900, in his Paris address, Hilbert responded regally: *'We hear within us the perpetual call: There is the problem. Seek its solution. You can find it by pure reason, for in mathematics there is no ignorabimus'*.

Cantor was also dismissive of infinitesimals in analysis. In the long, discursive, introduction to his *Grundlagen* paper in 1883 he addressed the question whether his sequence of transfinite cardinal numbers might help in the search for a consistent grounding of a theory of infinitesimals. Could these be regarded as 'finite' numbers which were distinct from rational or irrational numbers, but would lie between such numbers, or which, similarly, could be inserted between algebraic and transcendental numbers?

His answer was unequivocal: the theory of well-ordered sets would show such attempts, by various authors at the time, to be fallacious, resting on the one hand on confusion between the actual and potential infinite, and on dubious reasoning on the other. The description (by some philosophers) of the potential infinite as 'bad' infinity was unjustified, since it had frequently proved itself useful in mathematics and the natural sciences. To his knowledge, *all* uses of the infinite in real analysis could relate only to the potential infinite. He argued that any attempt to 'force' the infinitely small into the guise of an actual infinite would be purposeless. Even if such objects existed, or could be defined successfully, they could not stand in any immediate relation to normal quantities that were *becoming* infinitely small.

Yet the use of infinitesimals, if only as *façon de parler* used in textbooks, did not die out as Cantor may have wished. There continued to be serious attempts to construct number systems that included infinitesimals, notably in 1907 by the Austrian mathematician *Hans Hahn* (1879-1934), who constructed a non-Archimedean ordered field containing non-zero elements x such that, for all n in $\mathbb{N}$, $0 < |x| < \frac{1}{n}$. But there did not, as yet, seem to be a consistent way of exploiting such objects within real analysis.

This situation remained more or less intact for half a century. In 1958 a partial theory of infinitesimals was published by the German mathematicians *Curt Schmieden* (1905-1991) and *Detlef Laugwitz* (1932-2000). Their construction did not arouse much initial interest, as the number system they produced was not an ordered field. It was soon overshadowed by the publication in 1961 of an article by the logician *Abraham Robinson* (1918-1974), followed in 1966 by his ground-breaking volume *Non-Standard Analysis*, [39].

Robinson employed novel techniques from logic and model theory to construct an ordered field (necessarily non-Archimedean, of course) $^*\mathbb{R}$ – nowadays called the *hyperreals* – which extends the real number system $\mathbb{R}$ to include both infinite and infinitesimal numbers and is linked to real numbers in a way that makes applications to Real Analysis possible. This correspondence is based on a completely new concept, the *Transfer Principle*, which states (informally) that exactly the same formal statements (within symbolic logic, as described earlier) will hold in both $\mathbb{R}$ and $^*\mathbb{R}$. This may seem contradictory at first sight, since $^*\mathbb{R}$ contains infinitesimal elements, while $\mathbb{R}$ does not—the resolution of this conundrum lies in the precise description of the formal statements that are to be included in the Principle. In this sense the Transfer Principle might be regarded as a modern version of Leibniz' Continuity Principle. *Plus ça change!*

Versions of the hyperreals can be constructed in a variety of ways, not all equivalent. We will assume as *given* an ordered field that we denote by *$\mathbb{R}$, in which $\mathbb{R}$ may be regarded as an embedded subset (much as $\mathbb{Q}$ was taken to be embedded in $\mathbb{R}$ in **Chapter 7**). This extension extends to subsets of $\mathbb{R}$ as well as to real functions (i.e. between real numbers): $A \subset \mathbb{R}$ has a counterpart *A in *$\mathbb{R}$ such that $A = $*$A \cap \mathbb{R}$, and a function $f:A \longmapsto B$ extends to *f:*$A \longmapsto$ *B, so that f is the restriction of *f to A and the graph of *f is the extension of the graph of f.

The algebraic operations $(+, \times)$ and order relation $(<)$ similarly extend to *$\mathbb{R}$, as does the absolute value $(|.|)$ of a real number—for ease of notation, we will use these symbols also in *$\mathbb{R}$, without the asterisk.

In *$\mathbb{R}$ we can distinguish between three types of elements: $x \in$* $\mathbb{R}$ is

(i) *infinitesimal* if $|x| < \varepsilon$ for all $\varepsilon > 0$ belonging to $\mathbb{R}$,

(ii) *finite* if $|x| < r$ for some r in $\mathbb{R}$,

(iii) *infinite* (equivalently, *not* finite) if $|x| > r$ for all r in $\mathbb{R}$.

(Note that infinitesimals are also finite.)

A version of the Transfer Principle that ensures that all *first-order* properties of subsets and functions extend to *$\mathbb{R}$ is proved in Chapter 1 of **[1]**. In this setting, *completeness,* the key feature of $\mathbb{R}$, does *not* extend to *$\mathbb{R}$. For example, the set of all *finite* elements of *$\mathbb{R}$ cannot have a least upper bound in *$\mathbb{R}$: any upper bound of this set would be an infinite element, s say, hence $s - 1$ would still be an upper bound. This illustrates the need to take care when using the Transfer Principle.

For hyperreals x, y we write $x \approx y$ if their difference is infinitesimal, and say they are *infinitely close.* Clearly, a sum or product of infinitesimals is again infinitesimal, as are the inverse of an infinite element and the product of a finite with an infinitesimal element of *$\mathbb{R}$. So if $x \approx y$ and $u \approx v$ then $x + y \approx u + v$. If x, u are finite, then $xv \approx uv$.

Sums and products of finite elements are finite, as is the sum of a finite and an infinitesimal element. The inverse of a non-infinitesimal finite element is also finite. While the sum of two infinite elements of the same sign, and the product of any two infinite elements, are again infinite, products of infinite numbers with infinitesimals can belong to *any* of the three classes.

The following theorem ensures that any finite hyperreal is infinitely close to a unique real number.

Standard Part Theorem

If x is a finite hyperreal there is a unique real number $r \approx x$. So there is a unique infinitesimal δ such that $x = r + \delta$.

Proof: The set $A = \{a \in \mathbb{R} : a \leq x\}$ is bounded as a subset of $\mathbb{R}$, as x is finite. So $r = \sup A$ exists in $\mathbb{R}$. For any real $\varepsilon > 0, r - \varepsilon$ is not an upper bound of A, so we can find $a \in A$ with $r - \varepsilon < a \le x$, and so $r - x < \varepsilon$. As r is an upper bound of A, $r + \varepsilon$ is not in A, hence $x < r + \varepsilon$, so that $x - r < \varepsilon$. We have shown that $|x - r| < \varepsilon$ for all $\varepsilon > 0$ in $\mathbb{R}$, hence $\delta = x - r$ is infinitesimal. The uniqueness claim is obvious, as 0 is the only infinitesimal in $\mathbb{R}$.

We write $r = st(x)$ and call it the *standard part* of x. For finite elements, the standard part of their sum (product) is the sum (product) of their standard parts. If $x \le y$ in $^*\mathbb{R}$ then $st(x) \le st(y)$ in $\mathbb{R}$ (but not conversely). For finite x in $^*\mathbb{R} \backslash \mathbb{R}$ we see that $\delta = x - st(x) \ne 0$ is infinitesimal, and δ^{-1} is infinite. For infinite x, x^{-1} is a non-zero infinitesimal.

These preparations allow us to characterise convergence of sequences and continuity of functions very simply:

(a) for a real sequence (s_n) and $a \in \mathbb{R}$, $\lim_{n \to \infty} s_n = a$ is equivalent to:

$^*s_K \approx a$ for all infinite K in $^*\mathbb{N}$;

(b) if f is defined on $I = (a, b)$ in $\mathbb{R}$, then f is continuous at $c \in I$

if and only if $^*f(x) \approx^* f(c)$ for all $x \approx c$.

Both claims are proved by an appeal to the Transfer Principle, i.e. constructing a sentence in first-order logic that can be transferred. We only consider the second case (which is very close to Bolzano's definition of continuity):

Suppose $\lim_{x \to c} f(x) = f(c)$, fix any hyperreal $z \approx c$ and $\varepsilon > 0$. We need to show that $|^*f(z) -^* f(c)| < \varepsilon$.

We know that there is $\delta > 0$ such that the following (first-order) sentence holds in $\mathbb{R}$:

for all x, $|x - c| < \delta$ implies $|f(x) - f(c)| < \varepsilon$.

By Transfer, the corresponding sentence holds in $^*\mathbb{R}$:

for all X, $|X - c| < \delta$ implies $|^*f(X) -^* f(c)| < \varepsilon$.

Taking $X = z$ we see that $^*f(z) \approx^* f(c)$ whenever $z \approx c$.

For the converse implication, assume that $|^*f(z) -^* f(c)| \approx 0$

whenever $z \approx c$, and suppose that $\varepsilon > 0$ is given.

Taking any infinitesimal $Y > 0$ in $^*\mathbb{R}$, we have:

there exists $Y \in {}^*\mathbb{R}$ such that for all $X \in^* \mathbb{R}$, $|X - c| < Y$

implies $|^*f(X) -^* f(c)| < \varepsilon$.

By Transfer, the corresponding sentence holds in $\mathbb{R}$:

there exists $y \in \mathbb{R}$ such that for all $x \in \mathbb{R}$, $|x - c| < y$

implies $|f(x) - f(c)| < \varepsilon$.

We can take any such y as the required $\delta > 0$, hence $\lim_{x \to c} f(x) = f(c)$.

This elementary illustration should suffice to show that basic facts of Real Analysis can be recovered in an intuitively attractive manner—see, e.g. [1] for an account of this and a variety of advanced applications. In *MM* we outline the construction of a version of $^*\mathbb{R}$ reminiscent of Cantor's model for the reals.

Epilogue

I will leave matters there. Other extensions of the real number system have been proposed—see [28] for a striking example. By starting with counting and linking this to notions of the 'number line', we have seen how mathematicians have extended the number concept progressively over the centuries. Infinitesimals, it seems, are not as easily banished into the outer darkness as Weierstrass and others had supposed—but they had to change their clothes significantly in order to appear more respectable!

That said, what David Hume, who so clearly abhorred 'horn angles' (see the end of **Chapter 5**), might have thought about the infinitely many different orders of infinitesimals now made possible, must remain an open question. And it is easy to imagine how Kronecker (and, no doubt, Brouwer) might have reacted to the hyperreals. We might counter their concerns with Hadamard's confident assertion (quoted in Section 1) that *essential progress in mathematics* results from including notions which, for earlier generations, *'were "outside mathematics" because it was impossible to define them'*. And, after all, not even Plato's Olympian edicts stopped Archimedes from employing *neusis* constructions – nor, indeed, infinitesimal slices!

The present mathematical community has largely taken hyperreals in its stride, while seldom showing great interest in the details. One reason is the existence of a meta-theorem that maintains (roughly speaking) that any result which can be proved by nonstandard methods also has a 'standard' proof–which may well be rather longer, however! In this sense, the practitioners of nonstandard analysis appear to be closer to the current 'mainstream' than is the group at the opposite end of the spectrum, the constructivists, who not only reject infinitesimals, but also restrict the real numbers they accept to the numbers (essentially) definable in finitely many words.

Neither of these opposite poles has attracted more than a fairly small minority of practitioners to date. In both cases the 'entry fee' to participation, having to learn radically new techniques and adopt unfamiliar perspectives, may seem quite high to many researchers, trained as they usually are in techniques and subject matter still dominated by the groundwork laid in the late nineteenth century. Whether and how this may change only time will tell.

Let us therefore leave the last word to the venerable Sir Francis Bacon.

    https://doi.org/10.11647/OBP.0236.11

Etiam capillus unus habet umbram suam. (The smallest hair casts a shadow.)
Sir Francis Bacon, *Ornamenta Rationalia, or, Elegant Sentences*, 1625.

Bibliography

[1] L.O. Arkeryd, N.J. Cutland, C.W. Henson (eds), *Nonstandard Analysis: Theory and Applications*, NATO Science Series C, Springer Verlag, London, 2012, https://doi.org/10.1007/978-94-011-5544-1

[2] O. Becker, *Grundlagen der Mathematik in geschichtlicher Entwicklung*, Suhrkamp, Berlin, 1975

[3] M. Capinski, E. Kopp, *Measure, Integral and Probability*, 2nd ed. Springer Verlag, London, 2004, https://doi.org/10.1007/978-1-4471-0645-6

[4] A.-L. Cauchy, *Cours d'Analyse*, annotated translation by R.E. Bradley, C.E. Sandifer, Springer Verlag, New York, 2009, https://doi.org/10.1007/978-1-4419-0549-9

[5] J.H. Conway, R.K. Guy, *The Book of Numbers*, Copernicus, Springer Nature, New York, 1996, https://doi.org/10.1007/978-1-4612-4072-3

[6] L. Corry, *A Brief History of Numbers*, Oxford University Press, Oxford, 2015

[7] R. Courant, H. Robbins, *What is Mathematics?* Oxford University Press, Oxford, 1941

[8] R. Dedekind, *Stetigkeit und Irrationale Zahlen*, 3rd ed., Vieweg, Braunschweig, 1905

[9] R. Dedekind, *Was sind und was sollen die Zahlen?*, 4th ed., Braunschweig, 1918, reproduced by *Opera Platonis*, 2013.

[10] H.-D. Ebbinghaus, et al., *Numbers*, Springer Graduate Texts in Mathematics, 2nd ed., New York, 1998, https://doi.org/10.1007/978-1-4612-1005-4

[11] R. von Erhardt, E. von Erhardt-Siebold, 'Archimedes' Sand-Reckoner: Aristarchos and Copernicus', *Isis* 33 (5), 1942, 578-602, https://doi.org/10.1086/358623

[12] J. Fauvel and J. Gray, *The History of Mathematics - A Reader*, Macmillan, Basingstoke, 1987

[13] D. Fowler, *The Mathematics of Plato's Academy*, Oxford University Press, Oxford, 1987

[14] A.O. Gelfond, *Transcendental and Algebraic Numbers*, Dover, 1960

[15] D. Gillies, *Revolutions in Mathematics*, Clarendon Press, Oxford, 1992

[16] J. Grabiner, 'Who gave you the epsilon?' *American Mathematical Monthly 91*, 1983, 185-194, https://doi.org/10.2307/2975545

[17] I. Grattan-Guinness, *From the Calculus to Set Theory*, Duckworth, London, 1980

[18] G.H. Hardy and E.M. Wright, *An Introduction to the Theory of Numbers*, 5th ed., Oxford University Press, Oxford, 1979

[19] Sir Thomas L. Heath, *The Works of Archimedes*, Cambridge University Press, Cambridge, 1897, https://doi.org/10.1017/CBO9780511695124

[20] Sir Thomas L. Heath, *Aristarchus of Samos*, Cambridge University Press, Cambridge, 1897

[21] Sir Thomas L. Heath, *The Thirteen Books of Euclid's Elements*, Volumes I-III, issued by Green Lion Press, Santa Fe, New Mexico, 2017

[22] D. Hume, *An Enquiry Concerning Human Understanding*, reproduced by Project Gutenberg, https://www.gutenberg.org/ebooks/9662

[23] B. Jowett, *Theaetetus*, reproduced by Project Gutenberg, https://www.gutenberg.org/ebooks/1726

[24] E. Kasner, J.R. Newman, *Mathematics and the Imagination*, Gordon Bell, London, 1940

[25] V. J. Katz, *A History of Mathematics: an Introduction*, 3rd ed., Addison-Wesley, Boston, 2009

[26] W. Knorr, *The Evolution of the Euclidean Elements*, Springer Netherlands, Dordrecht, 1975, https://doi.org/10.1007/978-94-010-1754-1

[27] P. E. Kopp, *Analysis*, Edward Arnold, London, 1996

[28] D. E. Knuth, *Surreal Numbers*, Addison Wesley, Upper Saddle River, NJ, 1974

[29] I. Lakatos, *Proofs and Refutations*, Cambridge University Press, Cambridge, 1976

[30] E. Landau, *Foundations of Analysis*, 4th ed., Chelsea, London, 2001

[31] K. Menninger, *Number Words and Number Symbols: a Cultural History of Numbers*, (tr. P. Broneer), MIT Press, Cambridge, Mass., 1969

[32] G. H. Moore, *Zermelo's Axiom of Choice: Its Origins, Development and Influence*, Springer, New York, 1982, https://doi.org//10.1007/978-1-4613-9478-5

[33] E. Nagel, J.R. Newman, *Gődel's Proof*, revised, New York University Press, New York, 2001

[34] O. Neugebauer et al., *Mathematical Cuneiform Texts*, American Oriental Series 29, University of Michigan, 1945

[35] P. M. Neumann *The Mathematical Writings of Èvariste Galois*, Oxford University Press, Oxford, 2013

[36] K. Reich, *Carl Friedrich Gauss: 1777-1855*, Moos, Munich, 1977

[37] C. Reid, *Hilbert*, 3rd ed, Springer, Berlin, 1978

[38] K. Reidemeister, *Das Exakte Denken der Griechen*, Classen, Hamburg, 1949.

[39] A. Robinson, *Non-standard Analysis*, Princeton University Press, Princeton, 1966.

[40] W. Rudin, *Principles of Mathematical Analysis*, 3rd ed., McGraw-Hill, New York, 1964.

[41] G. Schubring, 'Bernard Bolzano—Not as unknown to his contemporaries as is commonly believed?', *Historia Mathematica* 20(1), 1993, 45-53, https://doi.org/10.1006/hmat.1993.1005

[42] L.E. Sigler, *Fibonacci's Liber Abaci*, Springer, New York, 2002, https://doi.org/10.1007/978-1-4613-0079-3

[43] E. Snapper, 'The Three Crises in Mathematics: Logicism, Intuitionism and Formalism'. *Mathematics Magazine* 52(4), September 1979, https://doi.org/10.1080/0025570X.1979.11976784

[44] I. Stewart, *Taming the Infinite*, Quercus, London, 2008.

[45] I. Vardi, *Archimedes, the Sand Reckoner*, https://www.lix.polytechnique.fr/Labo/Ilan.Vardi.sand_reckoner.ps

[46] Ernst Zermelo (ed.), *Georg Cantor, Gesammelte Abhandlungen mathematischen und philosophischen Inhalts*, Springer, Berlin, 1932.

Name Index

Index

infinite, 213
  null, 179
series
  divergent, 140
  Fourier, 212
  partial sum of a, 140
  representation of a function,
    134
  sum of a, 140, 190
set
  countably infinite
    (denumerable), 213
  derived, 212
  infinite, 213
  of first species (Cantor), 212
  proper subset of a, 153
  second species, 212
  simply ordered (Cantor), 224
  subset of a, 153
  theory, 152
  union, 153
  well-ordered, 224
sets
  equipotent, 219
short scale, 2
Stevin
  De Thiende, 79
  L'arithmetique, 80
successor, 201

tangent, 118, 121

telescoping sum, 118
Theodorus lesson, 36, 54
Theorem
  Pythagoras, 24
  Theaetetus', 36
theorem
  binomial, 121
  Bolzano-Weierstrass, 145
  extreme value, 145
  Gauss (roots of polynomials),
    104
  incompleteness, 244
  intermediate value, 145
  mean value, 147
  Rolle, 147
  Schroeder-Bernstein, 224
three famous problems
  doubling the cube, 70
  trisecting the angle, 72
Transfer Principle, 252
trigonometry, 74

variable, 139
Viete
  Canon mathematicus, 77
  In artem analyticam isagoge,
    78

Wallis
  Algebra, 98
  Arithmetica Infinitorum, 95
  infinite product formula, 95